EXPERIENCES NEAR DEATH

EXPERIENCES NEAR DEATH

Beyond Medicine and Religion

ALLAN KELLEHEAR

NEW YORK   OXFORD
OXFORD UNIVERSITY PRESS
1996

Oxford University Press

Oxford   New York
Athens   Auckland   Bangkok   Bombay
Calcutta   Cape Town   Dar es Salaam   Delhi
Florence   Hong Kong   Istanbul   Karachi
Kuala Lumpur   Madras   Madrid   Melbourne
Mexico City   Nairobi   Paris   Singapore
Taipei   Tokyo   Toronto

and associated companies in
Berlin   Ibadan

Published by Oxford University Press, Inc.,
198 Madison Avenue, New York, New York 10016

Oxford is a registered trademark of Oxford University Press

Library of Congress Cataloging-in-Publication Data
Kellehear, Allan, 1955–
Experiences near death : beyond medicine and religion / Allan Kellehear
p. cm.   Includes bibliographical references and index.
ISBN 0–19–509194–9
1. Near-death experiences.   2. Near-death experiences—Social aspects.
I. Title.
BF1045.N4K45   1996   95–3947
133.9'01'3—dc20

1 3 5 7 9 8 6 4 2

Printed in the United States of America
on acid-free paper

In memory of
Tom McDermott, O.M.I.

PREFACE

This book is a study of the near-death experience (NDE). I have chosen the title *Experiences Near Death* to signal the most important way in which the present work departs from others on this subject. I believe that it is valuable to enable the idea of "experiences" to encompass issues beyond the endlessly described features of the NDE. This is particularly important because these issues, as I will argue, have critical bearing on any human understanding of death, including the NDE. *All* experiences near death—euthanasia, abortion, NDEs, or ghosts—expose important divisions, anxieties, and uncertainties in any community. The NDE should be viewed against this complex and controversial background of experience.

Consider this. Death not only separates people in obvious ways, it also reveals the normally hidden or poorly recognized cultural and political agendas of various groups in any community. Experiences near death are common battlegrounds for competing social meanings and political visions. Contemporary dissatisfaction with religion or science, the problem of institutional authority, and the modern experience of depersonalization are some of the major anxieties and visions that people bring to their encounter with death. These, then, along with environmental and physiological considerations, should be included as a critical part of our experiences near death. But more important, these cultural and political contexts create the ways in which the NDE, or any other experience near death, is understood by both the lay and academic communities.

We do not see or think in totally neutral ways. Our ways of understanding are colored by an assortment of hopes and troubles, both private and public. Thus, the central question I am posing in this book is: what does the NDE, and the community and academic reactions to it, look like in various contexts? This question enables us to go beyond the popular medical and religious images of the NDE, to challenge their relevance and, at times, even their validity.

I have attempted to break away from the polarized and restricted parameters of religious or medical debate, a debate no one seems capable of winning, to address these social and cultural meanings of the NDE. I begin by examining the role of cultural institutions and social circumstances as primary influences on the appearance and shape of the NDE. I then examine the possible social and political meaning of both community and academic reactions to the NDE. I believe the NDE is more significant than an interesting footnote in the medical literature and far more important socially than its possible value as evidence for life after death.

Twenty years ago, when I was a university senior studying sociology, my old friend Tom McDermott gave me a copy of Raymond Moody's famous book on the NDE entitled *Life after Life*. He made me promise to read it, knowing full well that it was almost impossible to get me to read anything not on the monumental reading lists that the Sociology Department provided almost weekly. I was struck, as I continue to be today, by how extraordinarily different the NDE seemed to be from most things I had read or heard about. Even then, I did not feel that the significance—certainly any enduring personal and social significance of the NDE—was to be found in a theory of the afterlife or of hallucinations.

It has taken me those twenty years, working across a variety of academic fronts, from transcultural psychiatry to medical sociology, to slowly develop a modest, still incomplete picture of the social meaning of the NDE. For the first few years, I did not know what to think or say about the NDE as a sociologist. A year after giving me his unusual gift, my dear friend died from a long-standing heart condition. In the months before his death, as my most formidable and provocative intellectual sparring partner, Tom could not, to his chagrin, get me to weigh in with my usual force. Unfortunately, this was because, on this topic, I was uncharacteristically at a loss for explanation.

Tom never did hear the social and cultural theories about the NDE that a good Irish Catholic priest, such as himself, would have enjoyed debating until the wee hours. And although I was too slow for Tom, I did gradually accept the last challenge he threw my way. The book you are about to read is the product of those reflections and conversations that began with my old friend but then took me halfway around the world to discuss them with colleagues in diverse countries and equally diverse academic cultures.

In that context, I would like to thank La Trobe University in Melbourne for providing sabbatical leave so that I could work on this project full time. Ian Stevenson of the Division of Personality Studies at the University of Virginia, and Ian MacDonald at the Centre for the Study of Clinical Practice at Melbourne University's St Vincents' Hospital, both provided me with a place to work at different times, as well as the necessary intellectual support and stimulation needed for this work. Although their work is very different from mine, they were quick to share with me their friendship and their unwavering commitment to an interdisciplinary view of human problems. This has given at least one sociologist some shelter from the current social science uninterest in the NDE.

Some of the ideas in this book have appeared earlier, in American or British academic journals of medicine or thanatology. Chapter 2 was originally published in the *Journal of Nervous and Mental Disease* (1993: 181, 148–56), and Chapter 3 was originally published in *Social Science and Medicine* (1990: 31, 933–939). I am grateful for permission to reprint them here. Chapters 6 and 8 have appeared in volumes 10 (1991) and 12 (1993), respectively, of the *Journal of Near-Death Studies*, and I thank the editor for granting permission to reproduce these too.

I also owe a debt of gratitude to two long-standing supporters and influences on my work: Bruce Greyson at the University of Connecticut's Medical School and Carol Zaleski in the Religious Studies Department at Smith College. Both of these distinguished colleagues have actively promoted a more social and cultural view of the NDE through their own pioneering work. But well beyond those contributions, through our mutual friendship by mail, fax, E-mail, and the occasional surprise visit, both of these people have been important to the development of my own style of social theorizing.

Carl Becker at Kyoto University has also contributed to my thinking on the subject, in our private and published exchanges, in the United States and Japan. Ken Ring, in the Psychology Department at the University of Connecticut, must be one of the busiest men alive, but he has regularly found time to discuss and challenge my ideas on those occasions when we have met and in letters we have exchanged over the last couple of years.

I also owe a debt of gratitude to the members of the Health Sociology Research Group at La Trobe University, especially Evan Willis and Jeanne Daly, who have patiently allowed me to work out my various ideas over the years through our department seminar series. I thank Jeanne also for her patient and constructive reading of an earlier draft of the manuscript. To Beth Robertson, Merle Parker, and Elaine Young, I also extend my thanks for their kind attention in typing the manuscript.

Finally, I wish to thank Jan Fook, my colleague and companion throughout my odyssey into the NDE. Jan has been the person who hears all the final drafts and more than a few of the earlier ones that, thanks to her wise counsel and criticism, never leave the privacy of our home.

Bundoora, Australia   A.K.
February 1995

ACKNOWLEDGMENTS

The following extracts have been reprinted with permission:

Basil, R. (1991). "The Popular Appeal of the Near-Death Experience," *Jounal of Near-Death Studies* 10:64. Reprinted by permission of the author and Human Sciences Press.

Counts, D. (1983). "Near-Death and Out-of-Body Experiences in Melanesian Society," *Anabiosis* 3:119–20. Reprinted by permission of the author and Human Sciences Press.

Davis, J.C. (1981). *Utopia and the Ideal Society*, p. 21. Reprinted by permission of Cambridge University Press.

Eadie, B. (1992). *Embraced by the Light* (portions). Reprinted by permission of Gold Leaf Press.

Greyson, B., and Harris, B. (1987). "Clinical Approaches to the Near-Death Experiences," *Journal of Near-Death Studies* 6:42–43. Reprinted by permission of the authors and Human Sciences Press.

Kellehear, A. (1990). "The Near-Death Experience as Status Passage," *Social Science and Medicine* 31:933–39. Reprinted by kind permission from Elsevier Science Ltd.

Kellehear, A. (1993). "Culture, Biology and the Near-Death Experience," *Journal of Nervous and Mental Disease* 181: 148–156. Reprinted by permission of Williams & Wilkins.

Kellehear, A. (1991). "Near-Death Experiences and the Pursuit of the Ideal Society," *Journal of Near-Death Studies* 10:79–95. Reprinted by permission of the editor and Human Sciences Press.

Kellehear, A. (1993). "Death and Renewal in *The Velveteen Rabbit*: A Sociological Reading," *Journal of Near-Death Studies* 12:35–51. Reprinted by permission of the editor and Human Sciences Press.

Kellehear, A., and Heaven, P. (1989). "Community Attitudes Toward Near-Death Experiences: An Australian Study," *Journal of Near-Death Studies* 7:168. Reprinted by permission of the authors and Human Sciences Press.

Kellehear, A., Heaven, P., and Gao, J. (1990). "Community Attitudes Toward Near-Death Experiences: A Chinese Study," *Journal of Near-Death Studies* 8:166. Reprinted by permission of the authors and Human Sciences Press.

Midgley, M. (1992). *Science as Salvation: A Modern Myth and Its Meaning*, p. 58. Reprinted by permission of Routledge Publishers.

Moody, R., Jr. (1975). *Life after Life*, pp. 23–24. Reprinted by permission of Mockingbird Books, Inc.

Ring, K. (1991). "Amazing Grace: The Near-Death Experience as Compensatory Gift," *Journal of Near-Death Studies* 10:30–31. Reprinted by permission of the author and Human Sciences Press.

Ritchie, G. (1978). *Return from Tomorrow*. Reprinted by permission of Chosen Books.

Saavedra-Aquila, J.C., and Gomez-Jeria, J.S. (1989). "A Neurobiological Model of the Near-Death Experience," *Journal of Near-Death Studies* 7:210. Reprinted by permission of the author and Human Sciences Press.

CONTENTS

THE NEAR-DEATH EXPERIENCE

undoing the stereotypes

1

POPULAR IMAGES OF THE
NEAR-DEATH EXPERIENCE

When people think of the near-death experience (NDE), they commonly associate this experience with one or several mental images. The most frequent image in the popular mind is one in which:

A person is seriously ill or involved in an accident.
The heart stops beating and resuscitation procedures begin.
After the victim has been revived,
    he or she tells a remarkable story that involves
    a sensation of great peace,
    an out-of-body experience,
    the sensation of moving through a tunnel,
    life review and
    encounters with bright lights
    and/or deceased relatives or friends.

This image and its features are perhaps the most common ones, but when one begins to delve into the literature on the subject—to read only two or three books, perhaps—other images are evoked. In addition to the classical Western NDE image, it soon becomes apparent that there are NDEs from other countries. These are slight variants of the NDE image to which we are accustomed. Persons from India tend to see Indians in their NDEs; those from Melanesia tend to see villages. But much of the NDE literature assures us that despite these understandable differences, the basic features of the NDE are highly similar.

Another theme that emerges from these accounts is the problem these people seem to have when attempting to tell others. Doctors attempt to explain away their experience; relatives suspect a developing neurosis or the side effects of surgery or drugs; friends are embarrassed or patronizing. But despite these early and occasionally persistent reactions, the interest in NDE is strong and apparently growing.

Of great interest are NDE accounts that purport to see strange new worlds. George Ritchie's *Return from Tomorrow*[1] and, more recently, another best-seller by Betty Eadie entitled *Embraced by the Light*[2] relate the social features of this other world in great detail. These extended accounts, partly because of the detail provided and partly because of the religious language of their authors, convey more than any other accounts the heaven-like state of this other world. The religious language and imagery has attracted the usual responses—the criticism of humanist and psychoanalytic readers. NDEs, they say, are a further example of "pie in the sky," an imaginative product of the deep human need to deny the finality of death.

Furthermore, if you have read three books on the NDE, it's a good bet that one of them talks about the NDE as evidence for life after death. So, another well-known image of the NDE is one of transformation, spiritual values, and particularly the importance of love in all relations. There may even be a discussion of angels, reincarnation, and human paranormal powers. Much of this has been referred to as the "New Age" image of the NDE.

Finally, at least one of your chosen books probably contains a section on the medical explanations for the NDE. These explanations are usually picked over to demonstrate why they cannot explain the NDE. Occasionally, works such as parapsychologist Susan Blackmore's *Dying to Live* are devoted entirely to the task of arguing that these theories can indeed explain the NDE.[3]

These seven popular images of the NDE have three features in common:

1. Most observers treat the NDE as a single well-defined entity.
2. They all assume that there are only two ways to understand the NDE: religious—in terms of life after death; or medical—in terms of the dynamics and mechanics of the brain and/or the unconscious.

3. Most ignore the *social* dimension of the NDE, its popular attraction, and its academic explanations. Rarely considered is the fact that features of the NDE, as well as its popular appeal and its explanations, are dependent on social and political contexts.

My aim in this book is to challenge these attitudes toward the NDE by reexamining, and overturning the images that give rise to them. The NDE is not a single, monolithic entity always associated with death. It has been stereotyped well beyond the recognition of early observers such as Raymond Moody[4] and Ken Ring.[5] What began as a broad description of emotional content and cognitive images near death has become another set of stages or staged images near death.

The persistence of the stereotype goes beyond that of the popular tabloids and magazines; it is alive and well in both the medical and psychological literatures. In this way, the NDE has been seen either as falling within the province of medical sciences or as consisting of philosophic and religious speculations and theory. The choices are between a materialist and a non-materialist explanation—a two-party election, so to speak. The sociological and anthropological realities that might challenge this rigid polarization of opinions and debate have been nearly absent. Ideas about culture, environment, social change, or the social construction of realities were either ignored or sketched as background rather than foreground issues. Each year, yet another book made one of the following claims: NDEs are hallucinations, revelations, or revelations in hallucinations.

Finally, overreliance on medical images and psychological explanations of the NDE has permitted these perspectives to attempt to explain *other* social phenomena. People have been encouraged to explain the enormous popularity of the NDE in psychological, particularly negative psychological, terms. Regularly overlooked in those explanations is the fact that populations in any society are diverse, and that prior experiences and current anxieties in other areas of people's lives play a major role in any single response. This has meant that theories giving *single* reasons to account for the popularity of the NDE still dominate. But the stereotyping has not been confined to the NDE or its popularity.

The medical explanations have a fixed, menacing, and unassailable quality, especially for those who feel that the explanations are too reductionist or overzealous. Medical science explanations appear "hard," that is, difficult to understand and more objective than social or philosophical explanations. That stereotype hides many recent insights in the sociology of knowledge and in the history and philosophy of science. I intend to bring some of this material to bear on scientific explanations of the NDE. In believing that science is "hard," others have interrogated scientific explanations on their own intellectual terms. Science is questioned using science's rules of the game. The assumption is that the housekeeping is all in order, so the questions are directed to the conclusions. I do not assume that the homework has been done. We are all theorists about the world in general and the NDE in particular, and as C. Wright Mills once remarked, "Every cobbler thinks that leather is the only thing."[6] Scientists are no exception in believing that scientific explanations are to be preferred. This book is no exception. But here, at least, is a different voice. This book will reexamine popular images of the NDE, as well as theories about its popularity and its major explanations. I am asking you to think again: What does the NDE, and the community and academic reaction, look like when we place all of these in a broader sociological and political context?

Let us consider this question by reflecting on the seven popular images of the NDE in turn, beginning with the non-Western image.

The Non-Western Near-Death Experience

It took nearly ten years for non-Western accounts of the NDE to creep into the medical or thanatology academic literature. When these accounts finally appeared, they came from a variety of regions: China, India, western New Britain, and Native American cultures, among others. There was disagreement about the significance of these accounts. Some commentators felt that these accounts proved that NDEs were culturally determined hallucinations. Others felt that it would be usual—if there was an afterlife—for dying people to rejoin the dead members of their society. Therefore, similarities should not be surprising. Following is the account of one well-known non-Western NDE (from Melanesia) documented by the anthropologist Dorothy Counts.

ANDREW

Andrew is a young man who lives in the Anem-speaking interior village of Bolo. His death occurred at the small hamlet of Vuvu, which is located about a kilometre from Bolo. His experience, which was well known throughout the Kaliai area, was thought to be especially remarkable because during his vision he saw the spirit of a woman whose death had occurred shortly after his and about which he could have had no knowledge. His experience was also remarkable because, according to local definitions of death, he was dead for several hours. His kin had gathered, pigs for his first funeral feast had been killed and the meat was being prepared for division, and his grave had been dug before he returned to life. Shortly after Andrew's death occurred at Vuvu, the wife of the lay minister of Bolo took some food to the young boys residing in the Bolo men's house. She was returning to her own house when she collapsed and died in the village square. She did not recover and was buried the next day. It is this woman whom Andrew reports meeting during his near-death experience. After Andrew recovered from his illness, one of his legs was withered, and he now walks only with the aid of crutches.

The day I died I was very sick and was sleeping in my house. I died at noon [when the sun is high] and came back at six o'clock that afternoon [at dusk]. At the time I died there was a woman who hadn't died. She cooked food and distributed it. But when I died my spirit met hers on the road.

When I died everything was dark, but I went through a field of flowers and when I came out everything was clear. I walked on along the road and came to a fork where there were two men standing, one on either road. Each of them told me to come that way. I didn't have time to think about it, so I followed one of them.

The man took my hand and we entered a village. There we found a long ladder that led up into a house. We climbed the ladder but when we got to the top I heard a voice saying, "It isn't time for you to come. Stay there. I'll send a group of people to take you back." I heard his voice, but I couldn't see his face or his body.

I walked around trying to see him, but I couldn't. But I saw the dead woman that I had met on the road, I saw her leave me. I wanted to call out, "Hey, come back!" but I couldn't, for this house turned in a circle. I couldn't see the man who talked to me, but I did see children lying [on platforms] over the doors and windows. As I was walking around, trying to see everything, they took hold of me and took me back down the steps. I wanted to go back to the house, but I couldn't because it turned and I realized that it was not on posts. It was just hanging there in the air, turning around as if it were on an axle. If I wanted to go to the door, the house would turn and there would be another part of the house where I was standing.

There were all kinds of things inside this house, and I wanted to see them all. There were some men working with steel, and some men building ships, and another group of men building cars. I was standing staring when this man said, "It's not time for you to be here. Your time is yet to come. I'll send some people to take you back. You cannot stay. This woman you saw coming here, it was her time and she must stay. But you must go back."

I was to come back, but there was no road for me to follow, so the voice said, "Let him go down." Then there was a beam of light and I walked along it. I walked down the steps, and when I turned to look there was nothing but forest. I stood there and thought, "If they have started mourning for me, I won't go 'because the voice said 'Stand there and listen. If there is no mourning and no dogs howling, you go back. But if there is mourning you come back." '

So I walked along the beam of light, through the forest and along a narrow path. I came back to my house and re-entered my body and was alive again. I got up and told my father of my experience, for he didn't realize what had happened. I died at noon and came back at six o'clock. I spent a long time wandering around this house before they sent me back.

Q: When you were a spirit, did you see your body?
A: No, I didn't see my body, I just came back, and when I got up I was well and told everyone what had happened.

Q: Were you sorry or happy to come back?
A: I wanted to go back there. It's a happy place, and I wanted to go back, but I couldn't. See how my leg is crippled here.[7]

The Non-Western NDEs vary in different ways from the Western NDE allegedly arising from medical resuscitation. In the foregoing account, note the interesting differences: There is no tunnel or out-of-body experience. There does not appear to be any review of one's life. There is a notable absence of a living being of light so often reported by Western NDErs. However, there are similarities. In the above account, the NDE arises from a grave illness. There is also a meeting of deceased acquaintances and a peek into what appears to be a world beyond. Counts relates this NDE to two others, one of which does contain a life review. In still another account, Counts reports a vivid Melanesian dream that includes an out-of-body experience. Taken together, then, these Melanesian accounts seem to contain most of the elements of the Western NDE. Or do they?

No sensation of moving through a tunnel is mentioned, although a sensation of darkness and a field of flowers are reported. The life review occurring in one of the other Melanesian accounts is unusual since there is no precontact notion of judgment in these Melanesian people. Counts speculates on the Catholic missionaries' influence on these experiences and images. In any case, although these accounts are fascinating, on their own they cannot give us a broad picture of cross-cultural patterns of similarities or differences. A systematic review has long been needed; this is attempted in Chapter 2. These examples of NDE are drawn from India, China, western New Britain, Guam, Native America, Australia, and New Zealand. Although some of these are single cases, a broad pattern of experiences does emerge. This pattern seems to indicate two things. First, important features of the NDE do vary insofar as we understand them from the medical resuscitation image. Second, the variations must be accounted for by examining the way certain societies emphasize or downplay certain cultural images and symbols.

Once again, the documentation of non-western NDEs indicates a reluctance on the part of many researchers to shift from the single Western image of the NDE to a broader cultural analysis. This reluctance is related to the attachment to and preference for a reductionist medical image of the NDE. However, how sound is it to associate the features of NDE solely with illness and resuscitation?

The Medical Resuscitation

The origins of the chest-pounding, resuscitation image of the NDE are traceable to Raymond Moody's famous composite description of the NDE. This has been reproduced in many books and articles, and is testimony to both the popularity and importance of the image for many social and medical commentators:

> *A man is dying and, as he reaches the point of greatest physical distress, he hears himself pronounced dead by his doctor. He begins to hear an uncomfortable noise, a loud ringing or buzzing, and at the same time feels himself moving very rapidly through a long dark tunnel. After this, he suddenly finds himself outside of his own physical body, but still in the immediate physical environment, and he sees his own body from a distance, as though he is a spectator. He watches the resuscitation attempt from this unusual vantage point and is in a state of emotional upheaval.*
>
> *After a while, he collects himself and becomes more accustomed to his odd condition. He notices that he still has a "body", but one of a very different nature and with very different powers from the physical body he has left behind. Soon other things begin to happen. Others come to meet and to help him. He glimpses the spirits*

of relatives and friends who have already died, and a loving, warm spirit of a kind
he has never encountered before—a being of light—appears before him. This being
asks him a question, nonverbally, to make him evaluate his life and helps him along
by showing him a panoramic, instantaneous playback of the major events of his life.
At some point he finds himself approaching some sort of barrier or border, apparently
representing the limit between earthly life and the next life. Yet, he finds that he
must go back to the earth, that the time for his death has not yet come. At this point
he resists, for by now he is taken up with his experiences in the afterlife and does
not want to return. He is overwhelmed by intense feelings of joy, love, and peace.
Despite his attitude, though, he somehow reunites with his physical body and lives.

Later he tries to tell others, but he has trouble doing so. In the first place, he can
find no human words adequate to describe these unearthly episodes. He also finds
that others scoff, so he stops telling other people. Still, the experience affects his life
profoundly, especially his views about death and its relationship to life.[8]

As you read this narrative, consider the features that I will now describe.
First, note that the NDE is connected to *dying* ("A man is dying"). This is
an observation supported by medical observation ("he hears himself pro-
nounced dead by his doctor") and is therefore considered a legitimate and
accurate judgment. This prevents these experiences from being associated
with the less reliable domain of mere personal or subjective judgment. In
other words, the patient did not merely imagine or dream that he was dying.
Further evidence of his confrontation with death is the image of people
attempting to revive him ("He watches the resuscitation attempt from this
unusual vantage point").

Next, the basic phenomenology of the NDE is established—loud ringing
noise, tunnel sensation, out-of-body experience, meetings with deceased and
unusual beings, life review, return, changed personal attitudes. These features are linked closely and without comparison to other similar experiences.
The features of the NDE are associated with medical events (illness, accident, or injury) and circumstances (surgery, resuscitation, or hospitals).
Given this setting, it is logical to look for medical explanations. This has led
to two developments beyond Moody's original intention and work. First,
the medical explanations have assumed that the experiences are abnormal.
This has led investigators to look for explanations in pathophysiology and
has facilitated a comparison with other abnormal activities such as temporal
lobe seizure activity and unusual neurotransmitter production in the brain.

In any case, the images described have led many medical and psychological
researchers to look for universally applicable physiological explanations *before*
checking the cross-cultural data. Furthermore, the social context of the
NDE, whether death might be anticipated by the patient (due to future

surgery or suicide attempts) or unanticipated (due to a sudden accident), took a back seat to the search for biological explanations. Overall, the NDE image presented here is a unified one. It is a single well-defined entity. Although Moody outlines his qualifications and reservations about the prevalence of each of the described features, the vignette he presented does not reflect this possible diversity. Although Moody, like others after him, considered the possibility of these experiences in nonmedical settings, the vignette privileges the medical image. And it is the chest-pounding resuscitation image of the NDE that has stayed in the popular imagination ever since. In Chapter 3 and again in the final chapter, I will show how understanding of an experience depends on the type of knowledge used to explain that experience. I will show that the features of the NDE have little to do with being clinically dead. Simply being in physical danger, but in good health and consciously aware, is enough to elicit many of these experiences.

The Community Reaction

One common theme that emerges from NDE accounts is the reaction of other people to the experiencer's story. In Western accounts, people recall the stigma and other interpersonal difficulties encountered in attempting to tell their story. Clinicians Bruce Greyson and Barbara Harris discuss the problem:

> Occasionally individuals who were totally unprepared to face an experience such as an NDE may doubt their mental stability, but may fear to discuss the NDE with friends or professionals, lest they be rejected, ridiculed, or regarded as psychotic or hysterical. Negative reactions of professionals when the experience is brought up may further alienate the NDEr and inhibit him or her from seeking help in understanding and integrating the experience.
>
> As many NDErs gradually adapt on their own to the experience and its effects, their changing values, attitudes, and interests strain relationships with family and friends who find it difficult to understand or adapt to the NDErs' new beliefs and behaviour. On the one hand, some experiencers find themselves ostracized from their families and friends, facing the fears of their significant others that they have come under the influence of evil forces, or that they have somehow become bewitched. On the other hand, media publicity about positive transformations following NDEs may also lead experiencers'

families and friends to place the NDEr on a pedestal and to expect unrealistic changes, to expect oracular or healing powers and superhuman patience and forgiveness. They are then disappointed and rejecting when the experiencer is unable to live up to his or her assigned new role as a living saint.

Specific intrapsychic problems brought about by the NDE include continuing anger or depression at the return from the near-death state, and difficulty accepting the return, with what some NDErs refer to as "re-entry problems" or "withdrawal symptoms"; difficulty reconciling the NDE with prior religious beliefs or with prior values and lifestyles; excessive self-identification with the experience, so that one thinks of oneself first and foremost as an "NDEr"; and the fear that the NDE may reflect mental instability, a concern that often can be surmounted only be redefining normality.

Interpersonal problems brought about by the NDE include a sense of exclusiveness or separation from those who have not had a similar experience; a pervasive fear of ridicule or rejection from others; difficulty in integrating attitude changes with the expectations of family and friends; inability to communicate to others the meaning and impact of the NDE; difficulty in maintaining customary life roles that no longer have the same significance after the experience; and accepting the limitations and deficits of human relationships in lieu of unconditional relationships and feelings experienced during the NDE.

Particular problems may face the individual who has had an unpleasant or frightening NDE, or who feels continuing anger or depression at having been revived from the near-death state. Likewise, unique difficulties may face those who have an NDE in childhood or arising out of a suicide attempt.

The response of a counsellor, therapist, or health professional to an NDEr may influence critically whether the NDE can be integrated and used as a stimulus for personal growth, or whether it is hidden away—but not forgotten—as a bizarre event that clashes with the individual's everyday life and may be regarded as a sign of mental instability.[9]

There are three interesting points to note about the above discussion. First is the emphasis on the problematic image of the NDE. People are generally surrounded by friends and relatives who have difficulty understanding the experience. Second, the problem of other people's lack of understanding is compounded by the poor or inappropriate social reactions of these others. Finally, the difficulties experienced by some NDErs are not enjoined and placed in the context of other adjustment experiences such as community reentry after war, imprisonment, shipwreck, or chronic illness and disablement.

The first and second issues, the pathologizing of the community reaction to the NDEr, comes from too ready acceptance of negative anecdotes from NDErs. Parallels can be found in other areas of death and dying studies. In the social research on cancer sufferers, a common professional and lay opinion is that people who are dying of cancer commonly experience stigma. However, stigma is a two-way experience. People may develop negative attitudes toward those dying of cancer by adopting patronizing thoughts. But as Gordon Allport once remarked, those who threaten to discriminate may not do so.[10] This qualification attached to attitudes may mean that such thoughts do not materialize into action. For cancer sufferers this can result in a non-experience or even in a positive experience. When I asked 100 people dying of cancer if they had noticed any negative attitude or behavior toward them at home or work, approximately 80 percent said "no."[11]

We do not know how typical the negative experience is for NDErs. Some of these experiences may be attributed to poor community information from the 1940s to the 1980s. However, some surveys suggest that belief in life after death, as well as supernatural and paranormal personal experiences, is quite widespread. This seems to contradict the assertion of community ignorance about the NDE. These issues are addressed in Chapters 4 and 5. In Chapter 4, we will review the survey evidence of community attitudes and reactions to the NDE and related experiences. In Chapter 5, we will review the historical and social reasons why the NDE has enjoyed so much attention—both positive and negative—over the last twenty years or so. Finally, the validity of comparing community reentry problems of the NDEr with other similar experiences is taken up in the last part of Chapter 3. There shipwrecked castaways are compared with NDEs in clinical circumstances.

Visions of Another World

The fourth image that dominates the popular imagination—indeed, seems to fire it brightly—is the intriguing image of a world beyond this one. Although it is uncommon in most NDE accounts, lengthy single accounts regularly turn up on the best-seller lists. The following account, sent to Kenneth Ring by a Canadian correspondent, is typical.

> About this time, while I was still marvelling over what I had seen, my friend suggested we might be on our way; and becoming restless myself, I agreed. IMMEDIATELY we arrived at another location, on a beautiful street. We appeared to be alone there, except for the street-sweeper, who was responsible

for the spotless condition of the place. Here again, the colours and textures were outside my experience; and the road and the sidewalks appeared to be paved in some kind of precious metal. The buildings appeared to be constructed of a translucent material. I felt prompted to talk to the street-sweeper, and congratulated him on his efforts. He said work was a joy to him, and he derived his pleasure from doing the best job he could at all times. This statement nonplussed me somewhat, for I had never been enthusiastic about what I considered menial tasks. This man appeared absolutely sincere, however; and I was very impressed by his industry and the obvious love and care he brought to his work. While I was thinking about this remarkable fellow, we relocated again, INSTANTLY.

This time we were audience to a choir of angels singing. Angels were totally outside my reality at the time, yet somehow I knew these beautiful beings to be angelic. They sang the most lovely and extraordinary music I had ever heard. They were identical, each equally beautiful. When their song was over, one of their number came forward to greet me. She was exquisite and I was mightily attracted, but I then realized my admiration could only be expressed in a wholly nonphysical manner, as to a little child. I was embarrassed by my error, but it did not matter. All was forgiven in this wonderful place.

Instantly we arrived in an art gallery. It contained the work of the great masters of all time and all places. The display was both classical and modern. Some of the great works seemed familiar. Others were unlike anything I had ever seen, indescribable. The beauty and form of the sculpture and paintings on display were beyond words. A lifetime could be spent in this place, but to see everything I needed to see during this visit, we must move on.

Next we materialized in a computer room. It was a place of great activity, yet peace prevailed. None of the stress of business was present, but prodigious work was accomplished. The people seemed familiar to me, like old friends. This was confusing, because I knew there to be present those who lived on earth still, and those who had passed on. Some of them I knew by name, others by reputation; and all had time for me, to teach me if ever I need help understanding. One of them was Albert Einstein, whom I had always admired greatly but distantly, and this great man took time away from his duties to encourage me. He asked me if I would care to operate the computer, which was very complex and beautiful and designed to guide the path of destinies. I was flattered, but felt incompetent and unsure of myself in the presence of such greatness. I told him I would like to try, but I was afraid of making a mistake. He laughed gently, and reassured me, saying that error was not possible in this place. Encouraged, I seemed instinctively to know how to operate this unusual machine, and waved my hands in a pattern over the large keyboard, rather like playing a piano without touching the keys. I knew instantly the task had been performed perfectly, and it had somehow been of great benefit to someone. I was suffused with the joy of a job well done. I would gladly spend eternity here at this rewarding work if only for the tremendous feeling of well-being I had experienced as a result.

We continued our tour and arrived at a library. It was a vast old traditional building, containing all of the wisdom of the ages, everything ever said or written. Room upon room, shelf upon shelf of books stretched away as far as the eye could see. By that time I had growing doubt I was destined to stay in this mysterious yet familiar place, even though I knew in my heart it was home. I had the uncomfortable feeling I must return soon to resume my life. My guide, for by now I thought of him as such, told me I must study and learn from the infinite array of wisdom before us. I was dismayed, and said there was no way I was capable of such a task. I was told to simply make a beginning, to do the best I could, and that would always be good enough. There was plenty of time.[12]

Too many people have read detailed stories similar to the one above and dismissed them as one person's version of heaven. This conclusion has fueled not one but two enduring images of the NDE.

1. *Afterlife imagery as a reason for the popularity of the NDE.* Because these stories occasionally feature Jesus or angels, they are said to be images of the afterlife. This pleases some and displeases others. Those in the latter category use that interpretation as evidence for their argument that the attracting power behind the NDE is the desire for "pie-in-the-sky" immortality. I have never been convinced that the simple prospect of immortality alone was enough to attract people to the NDE, assuming that these infrequently oc-curring vistas do dominate their imagination and feelings. If afterlife images alone had such power, perhaps all religious books would be best-sellers. It has always fascinated me that shrewd observers of stock market or political events, who emphasize the complex, multifaceted influences behind these phenomena, are happy to give just one reason for the fascination with the NDE. The fact that the NDE emerges, as it does, in a cultural milieu where anxieties about social change meet within a broad range of social institutions seems to interest very few. That reluctance or impatience with complexities surrounding the popularity of the NDE suggests a dismissive mood among adherents of one-reason theories. Chapter 5 begins by dismissing the dis-missers and exploring the possible cultural and critical reasons why people are drawn to the NDE—remembering that different parts of the NDE may attract different people and elicit different reactions.

2. *Afterlife imagery as images of heaven.* The preceding vignette of another world is quickly seized on by many writers as an account of "the world

beyond." As we have already discussed, such images are seen as images of the afterlife; what else could they be? However, there are two closer and more important observations to be made about this other world. First, these other-world images emerge from a story about a society quite different from our own. The story does contain a considerable amount of sociological detail, its brevity notwithstanding. There is evidence of work and occupations, and some general outlines of social organization and control. Second, the story itself is fantastic, that is, many of its features are fantasylike. This is not to say that the story is fictitious; I make no claim concerning its validity. However, some of the experiences (e.g., the mode of travel; the library of the wisdom of the ages) are inconceivable given our current understanding of natural laws and political organizations. In this way, the story resembles other fantastic stories (e.g., science fiction, utopian literature, legends). Of what social significance are these observations to an understanding of the NDE?

Chapter 6 will explain how the utopian images in the NDE transport the reader or listener to new worlds. These images present, social and political possibilities either not found or overlooked in their current social situation. People discover or rediscover social possibilities and ideals for life in their own societies. For example, many people enjoy travel tales and stories of high adventure but do not apply these in any practical or realistic way to themselves. But even if one cannot go to the balmy South Seas in reality, one can certainly plant some trees in the front yard and strive to have more recreation under them on weekends. In totalitarian societies, stories of democracy do not create it, but they nevertheless inspire some people to push for greater accountability in the limited terms existing in their own nation. This is because images of utopia *inspire,* and in the context of that inspiration they are able to transform personal values and social commitments.

Images of ideal societies in the NDE facilitate personal and social changes. That is their personal psychological power and their social significance and meaning. This is the argument developed in Chapter 6.

The Coming of Neuroscience

In seeming contrast to New Age theories about the NDE is the work conducted over the last twenty years by neuroscientists—the electricians of the medical and psychology worlds. Studying the patterns and activity of the brain—its anatomy, neurotransmitters, and hemostatic mechanism—

the neuroscientists' entry into the debate about the causes of the NDE has been dramatic. The contribution of neuroscience is notable for its contrast to community reactions. The image of neuroscience is abstract, its language difficult and inaccessible to the average layperson.

Its dismissal of religious explanation is aggressive and relentless. Its skepticism toward anything that is not empirical or laboratory based is uncompromising. A fairly typical example is quoted below. This is part of a neurophysiological theory of the NDE as advocated by Saavedra-Aguila and Gomez-Jeria (1989).

NEUROPEPTIDES, NEUROTRANSMITTERS, AND NEAR-DEATH EXPERIENCES

There is extensive evidence that some endogenous opioid peptides play a fundamental role in the regulation of excitability of the hippo-campus. Leu-5-enkephalin and met-5-enkephalin have excitatory actions in the hippocampal pyramidal cells, especially in the CA1 field, and B-endorphin has a naloxone-reversible seizurogenic effect (McGinty, Kanamatsu, Obio, and Hong, 1986), while its intraventricular administration can induce nonconvulsive epileptogenic actions on the spontaneous electroencephalographic pattern of rats, especially from limbic areas. These effects occur at doses with no analgesic effects (Henriksen, Bloom, McCoy, Ling, and Guillemin, 1978). In any case, opioid peptides can alter seizure sensitivity to different stimuli, probably through indirect effects on cholinergic, monoaminergic, and aminoacidergic functions of the limbic system (Gellman and McNamara, 1984; Woodbury, 1984). In the rat hippocampus, opioids appear to act via a disinhibitory mechanism in the dentate gyrus and in the CA1 field, probably on inhibitory aminobutyric acid (GABA) interneurons (Neumaier, Mailheau, and Chavkin, 1988).

These findings support the idea that changes in the normal balance of neuropeptides and neurotransmitters could contribute to abnormal limbic activity. It has also been shown that, during moderate stress and brain trauma, there is a liberation of endogenous peptides (Kelly, 1982; McIntosh, Fernyak, and Faden, 1986; Brooks, Burrin, Cheetaw, Hall, Yeo, and Williams 1988), which could explain the appearance of hallucinations in stressful situations in alert, nonpsychotic patients (Modai and Cygielman, 1986).

Therefore, in the stressful situation that always accompanies the NDE, there is a possibility of endogenous peptide liberation resulting in alterations of limbic activity (Leitner, 1984). This mechanism may contribute by itself to the NDE mechanism. . . . If endogenous peptide-induced changes in limbic activity do play a role in NDEs, then we could expect similar phenomena during situations of extreme psychological stress, such as kidnapping, torture, or panic attacks.[13]

It's quite easy to see why so many nonscience readers would find the above narrative off-putting. A story with characters such as "met-5-enkephalin" and with subplots such as the "naloxone-reversible seizurogenic effect" is designed to be read only by consenting adults. The claim—sometime implicit, sometimes explicit—that this work is value neutral (as opposed to the "biased" New Age explanations) is frequently present in these tracts. But how neutral is science? And in particular, how impartial is neuroscience, especially in explaining the NDE?

If we scan the above quote, ignoring the jargon, we can see words that are readily recognized: *abnormal, stress, brain trauma, hallucinations, kidnapping, torture, panic attacks.* Without knowing anything about neuroscience, the casual, naive reader could be forgiven for thinking the NDE is a medical condition. This is no accident; nor is it an impartial description of the physiology of the event. Science writing, like all writing, has rhetorical devices designed to persuade. What separates much science writing from other writing is that scientists, and quite a few nonscientists, believe that their writing is above all this. In the NDE debate, this makes New Age writing appear totally partisan and neuroscience, by contrast, morally and methodologically circumspect. This is not so. In Chapter 7 we will take a close look at the rhetoric and metaphors used in neuroscientific explanations of the NDE. We will see that not only is there a rhetorical presence but that it is a strong and arrogant one. We will examine some of the social and political reasons for this shrill academic voice. The roots of this rhetoric go deep in the continuing conflict of science with religion; in the resistance of some elements in the science community to postmodern developments in science; and finally, in recent challenges to the dominance of the medical profession and its theories about death.

The Psychoanalytic Dismissal

The view that NDEs, and the popular attraction to their imagery, are based on a denial of death represents one major attempt to impose a deterministic theory of human nature on the NDE. The concept of death introduced here represents part of the psychoanalytic view that death is, indeed must be, annihilating. Things can be knowable, and with great certainty. The Freudian materialist concept of death is apparently one of these. Because such an image of death must be threatening, denial must be our major response to this threat. Everywhere around us there is ample evidence for this denial:

> It is not the consciousness of death but the flight from death that distinguishes men from animals. From the times of the earliest cave man, who kept their dead alive by dying the bones red and burying them near the family hearth, down to the Hollywood funeral cult, the flight from death has been, as Unamuno said, the heart of all religion. Pyramids and skyscrapers—monuments more lasting than bronze—suggest how much of the world's 'economic' activity also is really a flight from death. If death is part of life, if there is a death instinct as well as a life (or sexual) instinct, man is in flight from his own sexuality. If death is a part of life, man represses his own death as he represses his own life.[14]

NDEs, then, are primarily defenses, or defense mechanisms, operating in the service of this need to deny. Faced with the prospect of annihilation, the mind turns to a last-minute side show to relax us before the lights of the theater go out. The underlying stimulus for our universal *denial* of death is a corresponding universal *fear* of death. However, there are other arguments, and we must remember that they are *arguments* and not dispassionate, culturally neutral observations.[15]

For example, there are ample suggestions from the anthropological literature that fears and beliefs about death vary as much as they do in any other area of human life, that is, according to culture and custom. Despite this cultural observation, intellectuals who favor the psychoanalytic principle of death denial continue to use this as a dismissal, as a way of not explaining the detail and context of NDE. Very few theorists have attempted to address the details of the NDE; not all have reviewed the cross-cultural NDE data that would indicate important variations;[16] most concentrate on a narrow band of near-death circumstances, mainly accidents, rather than on surgical or illness-related death and revival;[17] and for at least one author, such psychoanalyzing has lead to florid comparisons with birthing experiences,[18] roundly dismissed later by philosophy[19] and clinical[20] colleagues.

However, from a surprising and unexpected intellectual quarter, we continue to see this practice clearly: in psychoanalytic responses to children's writing, particularly in such writing involving a death scene. The examinations of these stories involve the discussion of psychoanalytic concepts as if these were theoretically unproblematic and culturally neutral in their assumptions. An excellent example of this interface can be found in some critical responses to the famous children's story *The Velveteen Rabbit: or How Toys Become Real.*[21]

In this story, the Velveteen Rabbit undergoes a death scene that exhibits most of the central features of a Western NDE. This "death" of the rabbit, which appears to some psychoanalytically minded critics as "no death," has attracted one or two lengthy criticisms, but as I will attempt to show in Chapter 8, this is done at the expense of ignoring the context of the story.

My task in Chapter 8 is to use this story, and the debate about its interpretation, as a case study of the intellectual and professional prejudices that are sometimes aimed, unfairly, at some of the paranormal or mystical images of the NDE. There *are* ways of responding to nonmaterialist interpretations aside from debunking and dismissing. In this context, I supply a simpler sociological interpretation of the rabbit's NDE that uses the manifest themes of the story as my guide. I criticize the psychoanalytic reading for its over-interpretation but also for what I argue to be the distortion of evidence for its interpretation. Psychoanalytic discussions of matters regarding death are often a closed discourse rather than an open one between possible meanings. This analysis is driven by a preconceived materialist notion of death. Such materialism is seen as superior, as theoretically and culturally neutral, instead of part of the reservoir of cultural meanings about death that are part of our different cultural traditions.

Crisis and Meaning

In the final chapter of this book, I suggest that the key to understanding the NDE is to examine this experience in its social and environmental contexts. NDEs are crises, much like the experience of bereavement or of being lost in the desert or at sea. When we see the NDE as a crisis, we start to see the psychological and social effects of strange circumstances. We begin to see and appreciate what our usual responses are to a major disruption of our taken-for-granted world and its meanings. When certainty deserts us, we do not always find ourselves alone and adrift.

Technically, this is not a new insight. Raymond Moody, in his first book on the NDE, recognized that the unusual features of the NDE are probably not confined to it. Regularly in the NDE literature, we are reminded that there are near-death-related or near-death-like phenomena. But this is like calling Baptist religious practices "Catholic-like or Catholic-related" phenomena because both groups pray and believe in God. In fact, Catholics and Baptists do have much in common and much that make them dissimilar. However, this is because they are *religious* institutions, not necessarily more or less "catholic" (or "baptist"). A sociological analysis of religion must develop concepts where by the experience of religious practice can be understood in all its human diversity and commonality.

Similarly, the NDE must be seen as merely one type of crisis experience that displays a wide range of individual and social features, from personal transformation to mystical visions and experiences. I will argue that only by heading in this direction can we begin to ask penetrating and socially useful questions beyond whether the NDE is evidence for life after death or whether these experiences are only hallucinations. Such questions about the universal experience of human crisis include the following: What particular features of a crisis allow one person to reconstruct and transform his or her life, while another person is overcome and destroyed by it? Why do some people undergo disruptive personal experiences, sometimes in frightful environments, but do not experience a crisis, while others plunge into a spiral of personal change? This book does not supply the answers to these questions but it does suggest, through illustration and argument, that this is a more valuable direction for our future inquiries.

Furthermore, by drawing the NDE back into the context of crisis in which it is best understood, we can begin to recognize that death has no monopoly on personal change, insight, wisdom, or the paranormal, however we may define each of these. Unusual circumstances bring unusual experiences, but these are not abnormal. However, the more stable are our lives, the more predictable and certain our habits, the more resistant and complacent become our interpretations of the world. Under these circumstances, the circumstances of much of the stable democracies in the West, the more shocking and disorienting crises such as death and loss become. Hence the publicity and anxiety surrounding the topic of death in societies such as ours.

And so, if this is an insight regularly mentioned but not developed at any length, one last set of questions remains. Why can't those whose business it is to develop explanations, such as these, do so? Why do our intellectuals continue to overidentify the NDE not with crisis and the social affairs of the world, but with death alone and the physical nature of the body? Why do the major intellectual attempts to understand the NDE continue to go in increasingly narrow and reductive directions, instead of toward broad concepts that embrace culture, context, and social meaning? These are important and final questions to ask if the NDE is to be taken seriously as a fertile ground to be studied as crisis. And crisis, as I will argue, is the most promising conceptual direction for the future, particularly if NDEs are to have any meaning beyond the usual mud-slinging matches between skeptics and believers of science and religion.

2

NEAR-DEATH EXPERIENCES
ACROSS CULTURES

NDEs are now well known in the Western world in both lay and academic/ clinical circles. Furthermore, the majority of the research in this area has been with white, Anglo-European respondents.[1] Not surprisingly, the highly similar cultural backgrounds of these populations have resulted in highly consistent clinical presentations. This has led to claims about their universality.[2] It has also attracted the interest of biological theorists from psychology and medicine. Psychologists Blackmore and Troscianko, for example, assume that the tunnel sensation so often reported in the NDE is a central characterizing feature of it.[3] They argue that the tunnel sensation is secondary to the cortical disinhibition that accompanies the closing physiological moments of death. Psychiatrists Noyes and Kletti report that life review is a common response to death and can be explained by the depersonalization processes that accompany the hyperarousal of the temporal lobe.[4] This conclusion echoes the early work of Butler, who argued that life review is a universal response to death prompted or stimulated by the unavoidable physiological deterioration associated with the aging process.[5]

In this chapter, I will argue that features such as the life review and the tunnel sensation in clinical presentations of the NDE are not universal, and therefore that the development and general application of biological theories of causation are somewhat premature. The preference for biological explanations over simpler cultural ones is regrettable, if only because sufficient transcultural material exists to warrant reverse priorities. A review of most of the case material in this area will demonstrate the importance of exploring cultural factors more thoroughly before turning to biological ones.

The chapter is organized in the following way. First, I will provide a preliminary discussion of the main features of the NDE as we have come to know them in the clinical literature. Then I will present non-Western case material. This review of case material from different cultures will illustrate the diversity of the clinical phenomenology of the NDE, particularly between areas dominated by historic religions (Christianity, Buddhism, etc.) and those characterized by primitive ones. Although the accounts from Melanesia, Micronesia, native America, aboriginal Australia, and Maori New Zealand total only twelve, the following point should be noted: Because of the global Westernization and modernization processes now underway, many of these cases may be historically unique. Because of these processes, accounts from hunter-gatherer and cultivating and herding peoples will be particularly difficult to collect in their traditional forms. This small but significant sample may therefore represent important empirical and historical evidence that certain Western features of the NDE are probably culture bound. Undoubtedly, substantial changes to these cultures in the direction of Westernization will progressively obscure these findings in the future, making discussion of them more important now. That discussion will follow the review section. Finally, I will discuss the qualifications and limitations of the review before concluding.

Near-Death Experiences in the West

In its most popularly understood form, the NDE is triggered by a physical crisis that brings the person close to death, usually accident, misadventure or illness. Certain social crises—for example, being a castaway after a shipwreck—may also produce the basic phenomenology of the NDE; I will discuss these in the next chapter. The sensations of peace and of being out of the body (OBE), so that one may observe the accident or resuscitation scene as an observer, is the first major feature of the NDE. This experience may be followed by the sensation of traveling into a darkness or void often described as a tunnel, at the end of which one may meet a bright "being of light." At this point a life review may commence, sweeping across major and minor events of one's life. This is an educational evaluation rather than a judgment. Some time afterward, one may enter a transcendent world where deceased relatives and friends are met in another society similar to that of the individual.[6] Not all NDErs experience these events. Indeed, clinicians Greyson[7] and Sabom[8] have distinguished two or three variations, some of which do not include an experience of darkness or of visiting another realm beyond death.

Nevertheless, despite common internal variations both in the order and in the extent of the events, the above images well represent the composite Anglo-European experience and the clinical understanding of the NDE. Research in this area also concludes that NDEs in children closely follow the adult pattern.[9] However, non-Western cases of NDE differ in important ways from their Anglo-European counterparts. In the following section, I review only the NDE phenomenology associated with illness for two reasons. First, these circumstances are the main ones clinicians are likely to encounter. Second, these circumstances are the main ones associated with the NDE in the popular imagination. This, then, is a good starting point to show that culture, rather than physical processes, seem to provide the critical influence in these experiences.

Furthermore, despite certain anthropological parallels between clinical NDE and, for example, the social phenomenology of shipwrecks and other rite-of-passage experiences, one important distinction should be borne in mind. Although it is clear that similar social experiences may evoke similar psychological processes, it does not follow that these have a singular or even an overlapping set of physiological correlates.

For example, reviewing one's life consciously has different physiological correlates from reviewing one's life as part of a clinical NDE while unconscious. In both cases these life reviews involve similar psychological processes, but each clearly involves different brain areas and their respective functions. Which if either of the underlying physiological correlates is causative or derivative of these psychological processes is not critical here. However, since I wish to discuss the NDE associated with illness, I will not complicate matters by conflating its phenomenology with anthropological parallels that may have a different underlying etiology. In this respect, assessing the universality of the life review or tunnel sensation across all types of human experiences is *not* the purpose of this chapter. Instead, the more modest question of their presence cross-culturally in clinical NDE is the focus of the following review.

Non-Western Near-Death Experiences

CHINA

Contemporary accounts of Chinese NDE in the academic literature are fragmentary and piecemeal. What we know about Chinese NDE comes to us primarily through the historical work of Carl Becker[10] and the more recent empirical work by Zhi-ying and Jian-Xun.[11] Becker reviewed three traditional biographical accounts of well-known Chinese monks who were important to the founding of pure land Buddhism. Each monk experienced a serious illness that resulted in either an NDE or a deathbed vision while still reasonably conscious. In each of these accounts, no tunnel experience is reported, although one person proceeded "through a void."[12] Neither the OBE experience nor the life review was mentioned in these accounts. Encountering other beings, usually religious figures, and observing supernatural environs, usually interpreted as the paradisical "pure land," are consistent throughout the three narratives. In a later work, Becker[13] provided a secondary analysis of the works of Ogasawara[14] and Lai.[15] Ogasawara documents about 20 accounts of deathbed visions, and Lai documents up to 100. Becker argues that the analysis of deathbed visions reveals features of the NDE that are strikingly parallel, an observation made earlier by the parapsychologists Osis and Haraldsson.[16] Once again, though, there is no report of a tunnel sensation. However, emerging from a "dark tubular calyx" is reported.[17] There is no report of an OBE, but life review is suggested by one person who saw all his "sinful deeds."[18] Once again, supernatural environs and beings of light are witnessed.

Physicians Feng Zhi-ying and Liu Jian-xun conducted a recent study of Chinese NDE.[19] They interviewed eighty-one survivors of the Tangshan earthquake of 1976 and found that thirty-two of them reported NDEs. Their survey suggested that most of the Western NDE phenomenology was also present in their sample. OBE, the tunnel sensation, a sensation of peace, life reviews, meeting deceased beings, and sighting of an unearthly realm of existence were all reported in their study. Unfortunately, they did not include descriptive cases that we can analyze for content; thus, observations about their data cannot be scrutinized further. For example, although Zhi-ying and Jian-xun assert that "a tunnel-like dark region" was reported by their respondents, this is, in fact, a response to a prior descriptive category offered to them. Thus, we are unsure whether the tunnel sensation is a *volunteered* descriptor for this part of their experience.

Finally, a recent survey some colleagues and I conducted in China indicates that the Chinese experience and/or understanding of the NDE may not be very different from the Anglo-European one.[20] Similar to Zhi-ying and Jian-xun, we presented a typical Anglo-European vignette of an NDE to a sample of 197 Chinese in Beijing. Twenty-six of these respondents claimed to have had an experience similar to the one described in that survey. Overall it seems that, from the historical and survey evidence available, the Chinese NDE may be very similar to the Anglo-European NDE.

INDIA

The first major report of NDEs from India came to us from the work of Osis and Haraldsson, who interviewed 704 Indian medical personnel about their experiences with the dying.[21] In this sample there were sixty-four reports of NDEs. The remaining reports concerned near-death visions. More recently and directly, Pasricha and Stevenson reported sixteen cases of NDE from India.[22] In the majority of cases (ten) the respondents were actually interviewed by the authors, while in most of the others a "firsthand informant" was interviewed. In later studies, by Pasricha, another twenty-nine cases of Indian NDE were uncovered.[23] In a total of forty-five cases, then, Pasricha, and Pasricha with Stevenson, found no evidence of a tunnel sensation. There was one case report of an OBE. A life review was regularly reported, but this took the form of a reading by others of the record of the percipient's life. The panoramic review commonly mentioned by Anglo-Europeans is not reported in this Indian sample. The reading of a person's record is a traditional Hindu belief that, according to the authors, is apparently widely held or known to the people of India. Finally, observing religious figures and deceased beings is part of these Indian NDE reports. These beings are observed in a supernatural world whose features resemble the traditional view of the "other realm."

The Indian NDE accounts collected by Pasricha and Stevenson do not seem to exhibit tunnel and OBE features. However, Osis and Haraldsson, in their interviews with Indian health personnel, found several reports of OBE in Indian patients near death. Blackmore claims to have found cases of tunnel sensation in Indian NDE in her survey of eight respondents.[24] However, on closer inspection, all three of those who supposedly reported tunnel sensation actually reported a sensation of darkness. One respondent agreed that her experience of darkness was "tunnel like" only after accepting this suggestion from Blackmore. This raises two important methodological problems.[25] First, the acceptance of one descriptor does not mean that the description offered is entirely satisfactory. Another, less geometric descriptor, such as an experience of twilight or night darkness, may also be accepted, even preferred, if this is offered as a choice. Second, because Blackmore recruited her respondents through an advertisement in an English newspaper in India (rather than a Hindi newspaper), her respondents were not typical of people from India. We are therefore unable to draw any conclusion on this subject from Blackmore's study.

Life review and observing a transcendent world in Indian NDEs have parallels with Anglo-European accounts. However, the figures observed in this world, deceased acquaintances aside for the moment, are those suggested by traditional Indian or Chinese mythology. Nevertheless, as Pasricha and Stevenson warn, social variations in another realm, if it exists, should be expected, just as they exist in our own world. The appearance of familiar cultural images may be psychological, but it may also be sociological and empirical. In other words, either projection may account for the visions or the visions may actually be observations of another empirical world that resembles the world of its "expatriate" inhabitants.

GUAM

School psychologist Green reported four cases of NDE among the Chammorro of Guam.[26] Two of these cases involved direct interviews conducted by Green. The other two cases were collected by a local man who was interested in the subject of NDEs.

Like the NDEs gathered in India and China, the Chammorro cases report visits to a paradisical place of gardenlike appearance. Here the NDEer is met by deceased beings, some of whom are relatives. Unlike the Indian and Chinese cases, however, OBEs are reported; the respondents recount flying "through the clouds" and making invisible visits to living relatives in America.[27] There is no mention of life review of any sort in these accounts or of any tunnel experience. Indeed, the transition from the ill and unconscious state to the OBE appearance is unexplained. Respondents suddenly find themselves flying through the sky or walking on a road. The emphasis of the narrative is on the social experiences while unconscious, that is, of meeting deceased relatives or experiencing a flying visit to living ones.

WESTERN NEW BRITAIN

Counts reported three cases of NDE among the Kaliai as part of her 1981 anthropological fieldwork.[28] "Andrew," the subject of the case cited at length in the previous chapter, was one of the three interviewees. Once again, other realms are visited and deceased relations and friends are met. The afterlife environment, as in previous accounts, has a strong physical and social resemblance to the usual world of the percipient. So far these two features of the NDE, encountering other worlds and deceased beings, is a steady, recurring feature of NDEs. As we shall see in other non-Western cases, this trend will continue.

There are two points to note about these particular Melanesian cases. The first is the single report of a life review; the second is the absence of an OBE or tunnel experience. However, the picture is somewhat more complex than these first impressions may suggest. Although one person reported a life review, this respondent stated that he also witnessed a review of someone else's life also—a sorcerer. This review was narrated by the NDEr as a visit to a place where sorcerers are placed "on trial." Each person stands on a series of magnetic "manhole covers." If these hold the person fast, so that others must assist him in freeing himself, then he is called to account. If his explanation is unsatisfactory or unforthcoming, a series of punishing events occurs, ending with burning by fire. This is an unusual account, for as Counts notes, "there was no pre-contact notion of judgement of the dead for their sins."[29]

However, Counts notes that the western New Britain area has been "missionized" by Catholicism since 1949. Many of the Kaliai are at least nominally Catholic, although traditional and Christian ideas often exist side by side. This may account for the life review in this case. This is not the first case of mixed cultural imagery in an NDE. Pasricha and Stevenson report an American follower of Sai Baba (an Indian holy man) who almost died in a hotel.[30] His NDE featured the Indian life review of having his life record read by others.

Although no OBE was reported in the Kaliai NDE reports, one case is reported by Counts that may be a vivid dream, hypnagogic imagery, or an OBE.[31] Its nature is difficult to discern in that account because no dead or sleeping body is observed, nor is a new body identified. An ability to see unusual sights and travel vast distances is connected with characteristics of the spirit world. The question of OBE among the Kaliai, then, must be left open. There is a possibility that interpretations of similar experiences by Westerners may favor an OBE explanation, while those of the Kaliai may not.

Finally, in no case was a tunnel experience identified. All informants report the early part of their NDE as walking on a road. However, in one case the NDE began in darkness, which gave way to a walk in a field of flowers. Only after this part of the experience did the walk continue onto a road.[32]

NATIVE AMERICA

Schorer reports two cases of NDE from native North Americans.[33] These accounts were identified from H.R. Schoolcraft's nineteenth-century work *Travels in the Central Portion of the Mississippi Valley*.[34] In these accounts, OBE and encountering other realms and deceased beings are reported. The other realm, as in all previous cases, is similar to the former world of the percipient.

Absent from these two accounts are any reference to a tunnel experience or a life review. Similar to the accounts from Guam and western New Britain, percipients emphasize their journey. The narrative is a series of tales about what happened to them after they discovered that they were dead. This pattern is repeated in the only account to appear from South America. Gomez-Jeria reports a single NDE account from the Mapuche people in Chile.[35] In this account, an old man named Fermin was considered by his family and friends to be dead for two days. When he finally woke, he reported visiting another realm.

> He said that all his dead acquaintances, his own parents, his children, his wife, and other children that he did not know were all in there. There was also a German gentleman reading and writing in big books. When the German saw him, he asked what he wanted.
> "I am following my son" said the old man.
> "What is his name?" asked the German gentleman
> "Francisco Leufuhue."
> He called the guard and ordered him to inform Francisco.

After passing through a series of noisy gates, Fermin is reunited with his son, who tells him that it is not his father's time.

> When the time comes, I myself shall go to the side of the house to look for you. Then you will come. Now, go away.[36]

Note again, in this account, the absence of tunnel sensation or life review. The NDE reported here contains a visit to other worlds and the meeting of deceased beings. Jeria asserts that the presence of the German gentleman is an indication that culture "contributes in part to shaping the content of mental experiences."[37] This may be true for certain structural elements, such as, for example, *the presence or absence of* review phenomena, but we must be cautious *in explaining even the smallest detail of NDEs in cultural terms lest our cultural analysis fall prey to reductionist tendencies.* Clearly, NDErs meet an assortment of social beings, and their prior experiences shape their *interpretation* of the identity, function, and meaning of these beings. Only if we have strong evidence that these NDE accounts are purely subjective, like dreams, can we link even small details of the NDE content to culture and biography. We do not, at this stage, have that type of evidence.

ABORIGINAL AUSTRALIA

An isolated account of an NDE among Australian aborigines has appeared in several ethnographies during this century. It is, by all accounts, an unusual story in aboriginal terms because it is not a mythical account that can be interpreted as part of the aboriginal "dream time." On the contrary, the most interesting feature of this story, of which there are several versions, is that it is an historically real account of a human being who visits the land of the dead.[38]

Lloyd Warner retells a version of the account as "Barnumbi—and the Island of the Dead" in *A Black Civilisation.*[39] More recently, the Berndts (1989) have reported the same story told to them as "Yawalngura dies twice?" According to the Berndts (p. 376), the story is now quite old and part of a long oral tradition.[40]

The account is a long one, so I will only summarize the main elements. Yawalngura was out gathering turtle eggs with his two wives. He ate some of the eggs, after which he lay down and "died." Later, his wives returned from their own search and found him dead. They returned his body to the main camp and with others built a mortuary platform for him. After this, Yawalngura revived and told others that he became curious about the land of the dead. He decided to build a canoe so that he could travel there to visit. This he did and set off on a journey lasting for several days and nights. Finally, he arrived at an island where he met traditional spirits (e.g., the Turtle Man Spirit) and deceased beings who recognized that he was alive and had to return. These spirits then danced for Yawalngura and gave him gifts, such as a Morning Star emblem and yams for his return journey.

> Yawalngura took those things which were given to him. All the spirit people danced at that special spring (well), and they told Yawalngura that he had to return: "You have to return, you're not dead properly; you've still got bones. You can come back to us when you die properly.[41]

Yawalngura returned and told others of his fantastic epic journey. "Two or three days afterwards," however, Yawalngura died again, "only this time he did so properly."[42]

In this account, deceased beings and a land of the dead are visited. Again, both the people and the place have traditional mythical qualities. However, no tunnel experience and no life review are mentioned. Although the OBE is regarded as common in aboriginal Australia, especially during sleep and dreaming, no OBE is mentioned in this NDE account. These are also features of the final non-Western account from New Zealand.

MAORI NEW ZEALAND

In an autobiographical exploration of New Zealand white culture and its encounter with the native Maori culture, King recounts a Maori NDE.[43] Nga was a Maori woman who encountered her first white person when she was "a girl just over school age."[44] A favorite story of Nga's was apparently one about the occasion when she believed she had died.

> I became seriously ill for the only time in my life. I became so ill that my spirit actually passed out of my body. My family believed I was dead because my breathing stopped. They took me to the marae, laid out my body and began to call people for the tangi. Meanwhile, in my spirit, I had hovered over my head then left the room and travelled northwards, towards the Tail of the Fish. I passed over the Waikato River, across the Manukau, over Ngati Whatua, Ngapuhi, Te Rarawe and Te Aupouri until at last I came to Te Rerenga Wairua, the leaping off place of spirits.[45]

At this sacred place she performed the ablutions expected of the departed. Ascending to a ledge, she gazed down at the entrance to the underworld. After performing a dance, she prepared to descend into the subterranean passage leading to the realm of the spirits. At this point, she was stopped by a voice who told her that that her time had not come and that she must return until called again. She then returned to her body and awoke to see her anxious living relatives.

In this Maori account, no mention of a tunnel is made; instead, Nga flies to the land of the dead after her OBE (e.g., "I had hovered over my head then left the room"). However, the story of Nga takes us to the entrance of a subterranean underworld, and this, had she traveled to it, may have constituted a tunnel experience. This subterranean passage is a common feature

Table 2-1 Summary of Non-Western NDE Features

Culture	Source	Cases	Tunnel	OBE	Life review	Other beings	Other world
China	Becker (1981, 1984); Zhi-ying and Jian-xun (1992)	100–180	$/_2$	/	/	/	/
India	Osis and Haraldsson (1977); Pasricha (1992, 1993); Pasricha and Stevenson (1986)	64 29 16	X	/	/	/	/
Western New Britain	Counts (1983)	3	X_1	X_1	$/_2$	/	/
Guam	Green (1984)	4	X	/	X	/	/
Native America	Schorer (1985–86); Gomez-Jeria (1993)	3	X	/	X	/	/
Australian Aborigine	Berndt (1989)	1	X_1	X	X	/	/
N.Z. Maori	King (1985)	1	X_1	/	X	/	/

Key: X, none reported; /, reported; X_1, conditional negative; $/_2$ conditional positive.

of some Pacific cultures[46] and may mean, for the purpose of this review, that a tunnel experience cannot be excluded. If the experience had lasted longer, perhaps Nga would have descended to that underworld place through the traditional dark passage. So the absence of a tunnel sensation must be seen as a conditional matter that may relate idiosyncratically to this single account. Nevertheless, another characteristic, such as the life review, is unequivocally missing in this account.

SUMMARY OF NON-WESTERN NDE FEATURES

Table 2-1 summarizes the preceding review of the non-Western NDE. It identifies features that seem cross-cultural, those that appear to be culture-specific, and those in which the question of universality remains ambiguous. In every case discussed, deceased or supernatural beings are encountered.

These are often met in another realm, variously described as the "land of the dead," the "island of the dead," the "pure land," and so on. Consistently, the other realm is a social world not dissimilar to the one the percipient is from. The major difference is that this world is often much more pleasant socially and physically. Clearly, the consistency of these reports from highly diverse cultures suggests that at least these two features of the NDE are indeed cross-cultural.

This distinction is less clear in the findings about the OBE. Some cultures, such as the New Zealand Maori, native American, and Chommorro, clearly experience some kind of OBE with their NDE. However, the Chinese do not report this feature. Finally, the Australian aborigines do not report an OBE, but OBEs are known in this culture. In fact, OBEs are known in the vast majority of cultures,[47] but these may not necessarily occur, or occur consistently, in NDEs in these cultures. The apparent randomness of the finding concerning OBEs makes conclusions about them in relation to the NDE ambiguous. On the basis of the present data, we are unable to judge whether these are cross-cultural or culture-specific features. However, clearer patterns emerge when the life reviews and tunnel experiences are examined in the different non-Western cases.

Life review is a definite feature of the Chinese and Indian NDE accounts. In at least one Chinese account one's "sinful deeds" are observed, and in several Indian accounts one's life record is read from a book. A life review is also noted in connection with a Melanesian account, but this is a conditional finding. The life review described in the Melanesian account contrasts sharply with traditional—that is, precontact—notions of death and judgment. This particular feature and case may be better explained as a function of Western influence. Because many of the Kaliai are under Catholic influence, this may have altered either the experience or the narrative. This possibility gains some support from the observation that judgment is not part of the precontact Melanesian beliefs about death.

Life review is absent from all other accounts from Guam, Native America, aboriginal Australia, and Maori New Zealand. In other words, except for one ambiguous case in Melanesia, all accounts from hunter-gatherer, primitive cultivator, and herdsman cultures do not exhibit the feature of life review. Accounts from India and China, however, definitely do exhibit this feature.

Finally, the tunnel experience does not seem to be a feature of most non-Western NDE accounts. The cases from Native America, Guam, and India give no indication of a tunnel experience or sensation. There is, however, some suggestion that a tunnel experience may occur to those NDErs if their experience is prolonged.

This attaches conditions to any conclusion about this finding in the New Zealand case, as already noted, because the Maori NDEr traveled via a subterranean entrance to the land of the dead. Had she traveled to that place, she might have experienced a dark tunnellike experience, and this would have occurred after her OBE—as it does in some Western accounts. In the Australian aboriginal account, much may have been lost in its development into an oral tradition. For example, the visit to the land of the dead occurred after the percipient revived rather than during the time he was unconscious and "dead." It is difficult to speculate about what message or meaning is intended in this interesting turn to the story. Perhaps, having been dead, that person inherited special privileges/powers as a result of the experience that allowed him to travel while "alive" to the land of the dead. Perhaps the account has simply been altered via its oral transmission through the normal passage of time and embellishment. In any case, the trip to the land of the dead reports a journey of successive "nights," which indicates the importance in the account of light and darkness. This theme is also important in other aspects of the account, particularly when a struggle takes place over a "star."

In the western New Britain account, most of the NDErs describe how they suddenly appeared on a road, but one account does note that the individual emerged from darkness into a field of flowers. However, he did not describe this darkness as a tunnel. And in China, there exists at least one report of emerging from a similar darkness described as a dark void or "dark, tubular calyx." This calyx, of course, is the throatlike part of a flower and complements the lotus imagery of much pure land Buddhist narrative. In any case, Zhi-ying and Jian-xun report several accounts of tunnel sensation in their study of Chinese NDEs.

Overall, then, the present review has revealed that the major cross-cultural features of the NDE appear to include encountering other beings and other realms on the brink of death. Life review and the tunnel experience seem to be culture-specific features. Life review seems to be a feature of Western, Chinese, and Indian NDE accounts. Cases collected from hunter-gatherer, primitive cultivator, and herdsmen societies do not exhibit this feature. The tunnel experience is not described in most non-Western accounts, though an experience of darkness of sorts is often reported. The present review has revealed no major pattern in reports about the OBE in non-Western NDE accounts, and therefore this finding must be viewed as inconclusive. I now turn to a discussion of these findings, focusing on the following question: How are we to account for these differences in the pattern of non-Western NDEs, comparing them with each other and with Western accounts?

Discussion

The universality of NDEs/visions, in which individuals purport to see new worlds and beings beyond death's door, is not a particularly new or interesting finding. In relation to this recurring feature of death or NDEs, there has been a long tradition of medical social science literature, as well as similar concerns in the humanities and in religious thought generally. The conclusion is either that these visions represent observations of another empirical reality or that these are simply hallucinations. As delusions or visions of a dying organism, the sight of deceased relatives in a world beyond death may reflect archetypal forms, as outlined by the famous psychoanalyst Carl Jung.[48] They may represent a major cultural source of myths, but these myths may reflect neuronal activities and patterns, thus representing something of a cerebral map or blueprint.[49] Others have argued that perhaps the belief in and desire for the afterlife represents a strong unconscious need to deny the annihilating reality of death.[50] I will not attempt to arbitrate on an issue such as this, a task that is clearly beyond the scope of this book. Suffice to say that the universality of human experiences such as these has created and will continue to generate fierce, complex debate. Culture-specific findings are somewhat more modest in their demand for explanation, and I now turn my attention to these.

Blackmore and Troscianko asks two questions about the tunnel feature so often reported in the Western NDE[51]: First, why is the tunnel so often a regular feature of the NDE? Second, why do other symbols not appear— for example, gates or doors? The present review has not found tunnel experiences in any of the non-Western case material. However, there were several mentions of darkness described as a void, a calyx, or simply darkness. This suggests that tunnel experiences are not cross-cultural but that a period of darkness may be. This darkness is then subject to culture-specific interpretations: a tunnel for Westerners, subterranean caverns for Melanesians, and so on. NDErs who do not report darkness may not view this aspect of the experience as an important part of their account or narrative. Because an account of an event is a social exchange based on mutual expectations about what is or is not important information, recall may sometimes be selective and shaped to the perceived requirements of the listener. This may be an important methodological point for a review such as this because the sample of cases is too small to allow us to gauge if darkness is important or unimportant in all regional accounts. On the other hand, unless appropriate images can be selected that may convey traditional meanings about darkness, no image at all may be chosen to explain the experience. The first question that Blackmore and Troscianko asked must now be reformulated as follows: Why is the frequently reported sensation of traveling through darkness by Western NDErs so often described as a tunnel experience? In other words, why do many Western NDErs choose the term *tunnel* to denote their experience of darkness?

The term *tunnel* has two major meanings in the English-speaking world, and both of them are relevant to the NDE.[52] In a literal vein, tunnels are shafts or structures similar to the inside of a chimney. They may resemble subterranean passages or they may denote tubes, pipes, or simply deep openings that channel partway into other structures. Drab argues that tunnels have specific properties that make them enclosed spaces whose length is greater than their diameter.[53] This definition emphasizes the technical but in doing so omits the metaphorical, representational, and symbolic. This would not be important except that, in social communication terms, dimensions such as these must be viewed as equally influential in a person's choice of words. This is even more true when one considers that the NDE is often described as ineffable, that is, beyond words. Excluding the symbolic meaning involves excluding the figurative meaning, and the figurative meaning may be the most critical.

As a figurative term, *tunnel* may also denote a period of prolonged suffering or difficulty, as embodied in such expressions as "light at the end of the tunnel." Tunnels are often viewed as experiences of darkness that lead to other experiences. This representational view is well grounded in common experiences of the Western workaday world. It may begin in childhood, as for example, with stories such as *Alice in Wonderland,* reading of Alice's long fall down a dark rabbit hole that signals the start of a journey into a wondrous and confusing world.[54] Or it may begin with accounts of Santa Claus, who is expected to appear at the bottom of the family's chimney every December. From a child's kaleidoscope to the adult's experiences of gazing through telescopes, microscopes, and binoculars, Western people have grown accustomed to seeing strange new worlds through the dimness of tunnels. Through dark tunnels, and in the light that appears at the end of them, people leave the ordinary momentarily to experience the strange and unfamiliar.

Furthermore, tunnels are common images for the idea of transition, of traversing from one side to another. Drab objects to this notion of transition, arguing that often the tunnel in NDEs does not lead anywhere. However, this ignores the social fact that people may believe or *expect* the tunnel to lead them somewhere. Social experiences can rarely be understood in terms of concrete events separated from their interpretations, and these interpretive processes are constructed from attitudes, beliefs, expectations, and/or assumptions. If percipients are moving along in a shaft or space of darkness, some may choose a term that commonly denotes that experience. The tunnel is a symbol that, in Western industrial societies, is readily associated with that kind of experience. This brings us to Blackmore and Troscianko's second question: Why a tunnel and not other symbols such as a gate, door, or bridges?

However, gates, doors, bridges, and many other symbols do appear as images in the NDE, as Blackmore and Troscianko readily admit, but frequently alongside or after experiences of the tunnel. They rarely seem to substitute for the tunnel,[55] and this should be no mystery. The major and most obvious reason why gates or doors seldom substitute for tunnels is that NDErs are attempting to describe some kind of movement through darkness. It is the tunnel, therefore, rather than gates or doors, that best captures this experience. This is a further problem with a technical definition of tunnels that relies on shape as its primary characteristic. Shape reflects architecture rather than experience, but it is experience that is being described by NDErs. Because the experience is difficult to communicate, the descriptions will always be rich in interpretations that lean more toward metaphor than toward measurement.

The life review is the second feature of the Western NDE that seems to be limited in other cultures. Indian and Chinese NDEs seem to exhibit features of the life review, but NDEs in Australia, the Pacific, and Native North America do not. This might be explained by the scarcity of cases from these areas. A larger sample from these areas might turn up NDEs with life reviews. There is, however, much in the medical and anthropological literature to suggest that this will not be the case.

Butler argues that life review is always something of an identity search.[56] He uses the mirror as an example of an identity search metaphor that recurs in Western literature. In the Narcissus myth, the story of Snow White, stories from the Arabian nights, and also in the preoccupation of adolescents and the aged, the mirror reveals the face, the life, and the person. The mirror and the life review

> serves the self and its continuity; it entertains us; it shames us; it pains us. Memory can tell us our origins; it can be explanatory and it can deceive.[57]

This sense of self as interior, as inwardly responsible, driven and reflective from within, is a social construction of identity recently born in the development of what Bellah calls *historic religions*.[58] In historic religions, such as Buddhism, Christianity, Islam, and Hinduism, two worlds exist: the material and the divine. The self and the material world are devalued. Human nature is flawed and in need of rehabilitation and redemption, while the material world is mere illusion. The responsibility for rehabilitation lies in the action of the self.

In primitive[59] and archaic religions such as those of Native Americans, Australian aborigines, and many Pacific cultures, the distinction between self and the world is less explicit. "Mind" as a store for social experience is not paramount, for experience is also drawn from the animistic world of animals, vegetation, rocks, landforms, and climate. The mythological and actual worlds are not sharply separate but heavily overlapping.[60] Individuals are no more responsible than the world. Anxiety, guilt, and responsibility are in-the-world properties or characteristics, not located purely within the private orbit of an individual's makeup.

Roheim observed that the psychology of the Australian aborigines, for example, was based on much less internalizing of social sanctions.[61] They have a very good opinion of themselves, are easygoing, and fear the social consequences of transgression much more than private guilt, remorse, or anxiety. Laws are obeyed because they fear being caught.

Roheim discusses what happened when Christian missionaries began their proselytizing practices among the Aranda aborigines.[62] In their religious wisdom and magnanimity of those days, white missionaries promptly observed that the Arunda were all basically sinful and wicked and needed forgiveness before God. In response to this rather novel view of themselves, the Arunda retorted with understandable indignation. "Arunda inkaraka mara," they declared; the Arunda are all good. These were hardly a people who would seek a life review in evaluative terms or be impressed by a biographical review of their individual deeds.

So, although Indian and Chinese cultures may appear quite different from Anglo-European ones, in terms of social custom and language they are broadly similar in terms of religious development. Historic religion, whether Hinduism or Christianity, emphasizes the cultivation and development of the moral self vis-à-vis the divine world and its demands. Historic religions actively appeal to the notion of individual conscience. And conscience places great importance on past thought and action in the process of self-evaluation. As Weber has reminded us, when belief in spirits is transformed into belief in god, "transgression against the will of god is an ethical sin which burdens the conscience, quite apart from its direct results."[63] Since these religions link death with conscience, and conscience with identity after death, it is little wonder that some kind of life review takes place in near-death circumstances among people from these cultures. In a different world view altogether, members of aboriginal and Pacific cultures may not review their past personal lives in search of sense of identity. As mentioned earlier, the store of social experience is contained not within the self, but rather in the animistic and communal life of the physical and social world. There is probably little private use or function for a life review in individuals from this type of society.

Limitations

There are several limitations in the preceding review, making any conclusions drawn from this discussion tentative and subject to further investigation. All the cases were collected by different researchers, using different methods, sensitive to different aspects of the NDE, in highly different societies. While Becker, for example, noted the issue of life changes following an NDE, this played no part of the accounts of Pasricha and Stevenson and many others.

Language translation is also a problem. Not all words or phrases have an English equivalent. Indeed, not all social experiences are translatable, particularly outside of their contexts. Even Western NDErs struggle to find words in their own language to describe their experiences. A further complication is the lack of a widely agreed-upon definition of the NDE, an issue I will explore further in the next chapter.

The features of the NDE presented in this chapter are drawn from descriptive conventions of writers primarily from psychology and medicine. For some of these writers, features such as time distortion or subsequent life changes are an important characteristic feature of the NDE; for others, they are not. Furthermore, as I will argue in the following chapter, social circumstances play an influential role in the appearance or absence of certain features of the NDE. I will suggest that the social experiences of the person are critical to understanding psychological states underlying the NDE.

The current review of cross-cultural case material, however, strengthens the association between the social and psychophysical and suggests even broader links. For example, because tunnel symbolism, life review, the meeting of supernatural beings, and time distortion may occur separately or together in a variety of nonillness contexts, there is a critical need for cross-cultural research linking and comparing these to clinical presentations of the NDE. This will help us to arbitrate systematically and empirically on the difficult issue of whether similar social and psychological processes have similar or separate physiological etiologies. It will also invite discussions about whether the clinical phenomenology of NDE is merely part, a subset of experiences as it were, of the general social and psychological features of crisis and transition in ordinary human affairs. The current review identifies cultural differences in the clinical presentation of NDE particularly in regard to the tunnel sensation and life review. But it does not claim that tunnel symbolism and life review do not exist in other contexts for each of the cultures discussed. The question of the cultural factors that promote these processes in the psychological life of the individual in one circumstance (e.g., dreams) but not in another (e.g., near death) is still open.

The above limitations must not be thought to subtract from the current analysis or conclusion. On the contrary, they underscore the main purpose of the review itself. The review of cross-cultural clinical NDE material demonstrates that physiological theories of the NDE that do not incorporate a cultural theory of their context are methodologically much weaker for that omission.

conclusion

When people are very ill and near death, they may frequently report visions of another world. Just as commonly, this is said to be the world beyond death, a glimpse of the afterlife, a place where former friends and family members may be found. These experiences have a long history and appear in most cultures. They seem to be cross-cultural phenomena, part of the store of human possibilities that are not dependent on local influences of culture and socialization.

Other features associated with these visions in Western accounts have had a deep and dominating impact on the way we have thought about psychological processes in human beings at the brink of death. The clinical literature on tunnel and life review experiences has assumed a universality that seemed to point to some biological quality or character in the psychological construction of the self.

But, as I have argued, it seems that tunnel and life review experiences may be features of a certain type of human development in culture and psyche. In Western reports of a tunnel experience near death, this development materializes in the common shapes and symbols mentally associated with modernity and technology. If this conclusion has any validity, we should see the NDE darkness described as tunnellike mostly in modern, industrialized countries such as the United States and Great Britain. We should observe this description used far less often in China or India and rarely employed in hunter-gatherer societies. Our current observations do seem to support this cultural explanation. In the case of life review, religions such as Hinduism, Buddhism, and Christianity have cultivated an ethic of personal responsibility and conscience. This may be the chief influence on the evaluative style of mental processes near death. Life review in people with this cultural orientation is part of the general social and psychological process of identity formation. This is a task of ongoing importance in cultures with little or no regard for tribe or totem.

The role of culture in the social construction of the psyche is not fully appreciated by reference to child development in any single culture. Nor is this role given broader treatment by merely traveling further back in our history. A study of Hieronymous Bosch's sixteenth-century depiction of souls gravitating toward a tunnel on their way to the beyond is, however fertile and powerful an image, still one that is nevertheless culture bound in its significance. More cross-cultural studies, both empirical and archival, need to be undertaken. This work is necessary if we are to achieve greater validity in our understanding of human experiences near death. Only when we are more confident about the culture-specific elements will we then fully appreciate the size and nature of the task in explaining the universal features that remain.

3

UNUSUAL CIRCUMSTANCES, UNUSUAL EXPERIENCES

In this chapter, I want to suggest that NDEs are social phenomena. We have seen how features of the NDE vary across cultures, but here, rather than refer to broad cultural influences, I want to demonstrate that it is perceived social circumstances that play the crucial role in creating the experience itself. In the following discussion, I shall attempt to show how changing social conditions lead to altered perceptions and experiences of the self and the world. By sketching a theory of social context of the NDE, one is able to see that, far from being abnormal experiences, NDEs are surprisingly common, normal responses to uncommon, unusual circumstances. For too long now, the debate about the NDE by medical and psychological workers has given this area of human experience the appearance of oddity. Some of this can be traced to the suggestion of journalists and some academics that the NDE may be a window on the afterlife. Other images of abnormality can be traced to some traditions of psychiatric and physiological discussions that view the NDE as an aberrant response to illness or injury. Nevertheless, these views distract from the broader task of understanding the NDE as a regularly occurring social experience encountered by many people in a variety of settings *other than illness.*

It should be clear from the outset that I make no judgment as to the ultimate physical or metaphysical explanation of the NDE. Like the sociologists Max Weber[1] and Emile Durkheim[2] on religion, I assume that an examination of the eschatological claims of religion or NDE is unnecessary to an assessment of their cultural meaning. My approach has been phenomenological. This means that I have examined the experience from the point of view of the experiencer. In this sense, it is immaterial to the analysis whether NDEs are the result of cerebral anoxia or actual glimpses of some afterlife environment.

In this chapter, I am principally concerned with the sense and meaning that the experiencer constructs from the experience rather than its possible causes. This pursuit of social meaning will necessitate locating the NDE within a context of other social experiences of a similar type. There are two reasons for doing this. First, I acknowledge the sociological axiom that the private meanings of individuals are best understood when placed in a cultural context. Second, and flowing from the first point, NDEs should be seen as a member of a group of social phenomena because they have similar consequences for the experients and their social networks.

In the past twenty years, our understanding of NDEs has been medicalized. As we have seen, NDEs have been identified with medical[3] and psychological[4] debates and professions. The consequence of this has been the tendency to separate the NDE from other, nonillness social contexts. This process of decontextualizing has involved minimizing the similarities of the NDE to other social experiences in favor of comparing the NDE with similar medical and psychological phenomena. Furthermore, the search for psychomedical explanations has been a narrow one, often focusing on psychoneurological[5] and defensive mechanisms and explanations.[6] The emphasis has been on altered states of consciousness or physical functioning. The search for the sociological meaning has been slower and pursued with less vigor. This is regrettable because such searches add importantly to our understanding of the impact of life-threatening events on individuals and social systems. This impact, at both the personal and sociological levels, is not fully appreciated by confining our analysis to physical and psychophysical functioning. The approach taken in this chapter is a contribution to redressing this imbalance.

In attempting to develop an understanding of the social context of the NDE, I have organized this chapter in the following way. The first section will outline the psychological features of our medicalized view of the NDE and identify social features that can be extrapolated from these. The next section will introduce the sociological idea of *status passage* and show how dying and the NDE display characteristics of this process of status and identity change. The next section will then illustrate the essentially social nature of the NDE by examining and comparing the social experience of shipwrecked castaways. The role of status passage in shaping psychological reactions is explored in the next section. The final part of the discussion addresses the issue of whether our medicalized view of the NDE can legitimately be seen as related to other near-death encounters of a more social kind.

Psychological and Social Features of the Near-Death Experience

The medical and psychological view of the NDE, which I will refer to in this chapter as the *clinical NDE*, depicts a person with a serious illness or following an accident. During this experience, the person may perceive the following: being drawn through a dark tunnel; a sensation of being out of the body; a life review that may be facilitated by a "being of light"; the meeting of deceased acquaintances/relatives; and strong emotions of ecstasy, love, and peace, although fear and terror have also been reported.[7] Through successful resuscitation or inexplicable recovery, the person returns to the body. Another feature of the NDE, one that commonly occurs after the main experience, is a change in the values and attitudes of the percipient. Commonly, the NDEr reports less fear of death; less interest in formal religion; greater interest in spiritual values and lifestyles; a more accepting and tolerant attitude toward people in general; a desire for knowledge and, as a consequence, for education; and a desire to be of service/helpful to others.

If we use the psychological elements of the above description to identify the social ones, we observe three recurring features of a life-threatening crisis common to other transitional experiences of this kind. When we reexamine these features in broader social terms, the NDE unfolds in the following way. After an experience/period of sudden and unexpected separation, the person views his or her life and self as outsiders. In anthropological terms, the self temporarily takes leave of the "native's" point of view and adopts the viewpoint of the "other." During this process the person goes through a life review, sometimes with the assistance of another person. The reflections on this life and possible death can be disturbing or satisfactory but, in most cases, apparently transforming. Suddenly, and almost as unexpectedly as the original separation, the person rejoins his or her social network and the life belonging to that group. Afterward, a positive and humanitarian attitude to other people is a prominent feature of this person's thinking and subsequent behavior.

The three social features of the NDE, then, are (1) a sudden and unexpected separation, (2) a transition period involving the expectation of death, and (3) a sudden return to the original social group. Two sociological observations can be made about this description. First, these social features exhibit multiple characteristics of a status passage. Second, NDEs, when viewed in this way, are not unique social experiences but rather belong to a small but commonly occurring group of similar experiences. Let me begin with the first point. Consider the NDE as a status passage.

The Near-Death Experience as Status Passage

The sociologists Glaser and Strauss describe a status passage as consisting of social processes of transition from one part of the social system to another.[8] This frequently involves "a changed identity or sense of self, as well as changed behavior."[9] Glaser and Strauss speak of status as a "resting place for individuals." People occupy numerous positions at home, at work, and in other situations, and there is always the underlying assumption that these positions are temporary. They are temporary in the sense that no one holds a position forever, if for no other reason than that death will remove them. But death itself is only one reason, and not the most common reason, for changed status, the processes of which facilitate the social transformation of identity itself.

Important transitions such as adulthood, marriage, career advancement, and so on are much talked about and desirable. The regulation and control of deviance and inequality in society make other status passages such as imprisonment, unemployment, and social expulsion and rejection equally talked about and undesirable. The combination of illness and misadventure in the nature and social course of events makes passages to the status of patient, victim, appellant, and so on common and often unexpected experiences. All such transitions involve an experience of separation from the former status, a period of personal reorientation, and then an integration and recognition of the new status. By these simple social processes, people are allocated their individual roles and statuses by their communities and associations.

According to Glaser and Strauss, dying itself can be viewed as a status passage with the following characteristics.[10] Dying in contemporary European societies is usually unanticipated, and the social behavior itself is unprescribed by social custom or institution. Dying is generally undesirable and usually involuntary. The clarity of the signs of dying are not always clear to the caretakers or the dying person. Control over the passage of dying is problematic and uncertain for both caretakers and the dying person. Issues of reversibility (of recovery) and repeatability (relapse) are of ongoing concern until quite late in the passage.

The general description of dying as a status passage can, with little modification, be applied to the NDE. The NDE is usually unanticipated, that is, it is unexpected in the social and medical course of events. There is usually no social custom and institutional prescription for the NDE. There are isolated exceptions to this in the *Tibetan Book of the Dead* and in similar esoteric and unfamiliar sources.[11] The event that precipitates an NDE is usually regarded as undesirable and is usually involuntary. People who attempt suicide may be seen as exceptional. Signs of dying may not always be clear to the person who experiences sudden cardiac arrest but remains conscious. The experience of the NDE is not detectable by those attending the NDEr. Instead, the issue confronting these people is usually the more pressing one of resuscitation. Control over the NDE is uncertain. It is difficult to ascertain whether control resides with those who are resuscitating, or with the NDEr who desires to return, or with the being of light or deceased relatives who bade the NDEr to return because it is not his or her time. Variable though these factors are, it seems certain that control is always partial and depends on a variety of idiosyncratic factors. If one favors a medical view of causes, then these factors will be chemical and physiological. If one favors a psychological or religious view, then these factors will be the fickle decisions of the self or others on the "other side."

There is one way in which the NDE does not fit the usual pattern of status passage, and this makes the NDE status passage similar to dying as a status passage. Glaser and Strauss assert that one common hallmark of a status passage is the frequency and freedom of discussion surrounding the passage.[12] As mentioned earlier, passages ranging from adulthood and marriage to imprisonment and unemployment are much-talked-about transitions. This is less so, however, for dying as a status passage and still less so for the NDE as a status passage. In the case of dying, discussion may occur frequently among caretakers, such as doctors, nurses, and family members, but (when Glaser and Strauss wrote in 1971) seldom with the dying person.

For the dying person, dying as a status passage was a journey of "closed awareness." Although this situation has changed dramatically since Glaser and Strauss wrote, it is circumstantially true that NDEs are sudden events that rarely permit prior discussion.[13] But more than this, NDErs have long encountered a reluctance among professionals and some sections of the community to discuss their experiences.[14] NDEs have been status passages that have frequently gone unrecognized, especially in the 1960s and earlier. Instead of being viewed as transitions, they have been seen by many professionals as medical or psychological aberrations. NDEs have been stigmatized by this view. Alternatively, NDEs have been viewed as uncomfortable and inconvenient challenges to both materialist and religious images of death. They have also, then, been stigmatized by this view. Unlike most status passages that reflect the transition of people between established status structures within the social system, the NDE status passage has represented an event beyond the control and regulation of normal society. It has also represented potential criticism of that society by highlighting inadequacies in the usual system of control and regulation.

The NDE as a status passage, then, is unique for two important reasons. First, it is marginal to the usual social processes of mobility. In this sense, it is not a structural, normal, or usual path to social influence, value change, or identity formation. Second, and because of the first characteristic, this status passage, unlike all others, represents a powerful form of social criticism. But if clinical NDEs are the only type of this particular status passage, is this important in broad cultural terms? It is indeed important because the incidence of clinical NDEs is commensurate with a medical world where resuscitation has become common. It is also important because the clinical NDE is not the only type of status passage that conveys an expectation of dying. This is because the NDE as a social phenomenon is both broader and more frequent than one would first imagine. Using the social features of the NDE as a guide, we can observe other similar and frequently occurring NDEs.

Other Near-Death Experiences and the Status Passage

If we move away from the medical conception of the NDE as a psychological experience due to serious illness or accident, and if instead social features are used as a guide, the NDE can be found in other unpredictable social affairs.

Medical and psychological etiologies are less important in understanding the social experience. The social phenomenology is what counts for the experiencer and is also the most important criterion for identifying other *social NDEs*. The experience of being a castaway is a good example of another type of social NDE that can be viewed as inducing the same kind of status passage.

THE EXPERIENCE OF BEING A CASTAWAY

A shipwreck is an unscheduled, unanticipated event regarded as unusual in the normal course of travel. Although travel mishaps regularly occur, they are seen as uncommon and unfortunate deviations from the norm. There is usually no social custom or institutional prescription beyond general advice about who should occupy life boats if exceeded by the number of passengers. Although it is true that few large ocean liners are now shipwrecked, and those that do are often electronically equipped for fast rescue, many smaller craft commonly experience problems.

Many records of shipwreck exist. For example, Robertson reports his family's ordeal after whales rammed their yacht.[15] The Baileys documented their 117 days adrift after their yacht met the same fate.[16] In Australia recently, two more shipwrecks were reported by the occupants of small crafts. In one case, four crewmen of a sailing vessel were adrift near New Zealand for four months.[17] In another account, a couple were adrift for two weeks.[18] In these accounts, the ordeal occurs in a context of uncertainty and danger. Death is a distinct possibility. Reflection on life and death is an important feature of the experience. Culver noted in her account:

> That first day we talked a lot. I said I wanted to have a child before I got any older. . . . We also made plans about living abroad . . . work . . . where we would live. . . . [19]

Robertson recorded the following:

> We all felt a little sad and depressed at the prospect of our imminent demise especially Neil, who, I felt, could visualise more clearly the privations which lay ahead. . . . [20]

Robertson's fellow castaways, who included children, also wrote farewell letters to friends or relatives to store in the pockets of the raft.

In another account, some of the four crewmen came to relate their ordeal to a "being" outside the small group:

> Jim came to understand [that] someone up there was looking after us. The further on we went and learnt our lessons the more it became apparent.[21]

Robertson also noted this:

> Lyn is quite convinced that a spiritual presence assisted us in the storm. . . . [22]

The appearance or perception of this being is noteworthy in the castaway accounts because this "sensed presence" has also been associated with the separation of widowhood.[23] I make no comment on its possible origins. I simply make the broader sociological observation that perceived encounters with supernatural or deceased beings are not confined to instances of gravely ill and unconscious people so often associated with the clinical NDE. Such encounters are also frequently part of the broader conscious social experience of ordinary people.[24] This is especially the case when these people undergo the trauma of separation from cherished relationships because of death or, in the examples of these castaways, the experience of prolonged separation and stress.[25] Castaway accounts also reveal a new humanitarian attitude toward others:

> I'm a new man, I hope they appreciate it [the experience] and make other lives better for their experiences too.[26]

A woman who spent two weeks as a castaway with her partner recalled that they had spoken of NDEs. She spoke of her attitude toward life after her rescue this way:

> if we are saved, if we ever come back from the other side, we would devote part of our life on this Earth to helping others. Well, I think I've returned from something near the other side. . . . [27]

All of these cases reveal many of the elements commonly found in accounts of clinical NDE. There is a period of separation and a review of one's life—sometimes with the assistance of another. There are reports of a sensed presence. Emotions range from terror, fear, and despair to elation and euphoria, particularly when rescued. Transformation of attitudes and values is experienced—commonly in a positive, humanitarian direction. Rescue provides a sudden and sometimes difficult return to the normal social world.[28]

Now I do not wish to suggest that the above features of the social experience of shipwreck occur in every case. Nor are all the features of clinical NDE encountered even by the majority of those who have these experiences.[29] Some people who are shipwrecked may not review their lives; others may not sense a presence. Equally, not all NDErs experience life review or a being of light or meet deceased acquaintances. The nature of the circumstances themselves, and the length of time spent in this state, play some role in the felt intensity and evolution of the various social features. For example, the anticipation or expectation of dying, is an important influence on the types of features one might encounter in a clinical NDE.[30] Such variables may also prove important in the kinds of social and psychological experiences encountered by castaways.

The critical sociological point concerning the experience of shipwreck and clinical NDE is that both have identical features of status passage. Not only do the social processes of separation, transition, and return characterize the experience, but they do so in highly similar ways. Shipwreck and NDE are unscheduled. They are undesirable, involuntary, and unprescribed behaviorally. The event initially goes undetected, and although the loss is identified by potential rescuers as "missing persons or craft," the social experience of shipwreck for the castaways is not revealed by this identification. Control is an uncertain issue. It depends on the weather; one's own social and technological resources; proximity to fishing and passenger ship lanes; and the extent and comprehensiveness of search and rescue attempts. Like clinical NDE, a shipwreck provides criticism—de facto and sometimes actual. Why did Culver and her companion not have a long-range radio or an emergency locater to transmit? Why had the extensive search and rescue plan executed from New Zealand failed to find the boat belonging to the four crewmen who had radioed their last position? Explanations of clinical NDE and shipwreck are often incomplete. Blame and criticism focus on the instituted arrangement or the experient. The need for explanation and accountability is great, for in the case of clinical NDE, unusual medical or social events have taken place. In the case of shipwreck, the technology, the assumptions about travel, and perhaps the search and rescue strategies have revealed flaws.

In both cases, regulation of societal events has failed, and this failure has made a great emotional impact on the survivors. In this way, the context-dependent features of this breakdown may be seen as responsible for much of the private reactions. The social conditions of separation and privation that commonly accompany this kind of status passage are important in inducing these reactions. Importantly, these conditions, and therefore these reactions, are not confined to clinical examples of NDE.

The Personal Experience of Status Passage

Shipwreck is not the only experience that shares the status passage features of clinical NDE. Mining disasters[31] and mountaineering falls[32] also fall into this category and, in turn, may prompt and propel the experiencers through the social processes of transition to new values and identity.

But how do the social processes of status passage propel people toward new values and identity? Three characteristics of this status passage are important in creating the social experience of dying in the following way: First, unanticipated and involuntary separation creates a dramatic sense of loss of control, leading to an experience of social death.[33] Since most aspects of life are prescribed and controlled institutionally, their marked absence renders the experients marginalized and totally unsupported by their society. Second, if this experience is dramatic or prolonged, or both, the experience of social death will stimulate the psychological processes commensurate with an anticipation of physical death. Physical death, in turn, becomes a focal point of thoughts and emotion. Under these social conditions, the status passage moves the individuals forward from their usual collection of mainstream statuses (as parent, spouse, worker, etc.) to the marginal statuses of victim, dying person, castaway, and so on.

In this anticipation of death brought about by these social conditions, life review begins. Noyes regards this anticipation of death as the "chief prerequisite" of the psychological processes in accidents and drowning.[34] This seems less the case with clinical NDE, in which, as Greyson observes, anticipation of death leads to less cognitive experiences such as thought acceleration and life review.[35] The collection of cases that Greyson draws on for this particular conclusion, however, tend to be suicide attempts, exacerbations of chronic illness, or complications of surgery. In all of these cases, anticipation of death and hence life review may take place before physical death in the same way as for castaways. Serious chronic illness and surgery often lead to the expectation of complications and death, and these may reasonably lead the experients to review their lives as responses to these possibilities. Except for sudden and impulsive suicides, many people who attempt suicide may do so in the context of some kind of review of personality, circumstances, and life. So, although life review may be absent or less frequent in these cases of *clinical* NDE, they may not be if one takes a broader view of the NDE as a *social experience*. In this way, the NDE can be seen in a context that includes the social experiences before and leading up to the suicide or medical crisis.

The life review process has two important functions, psychological and sociological, each reinforcing the other in the process of reshaping identity and social values. First, the review is a climatic experience for the individual.[36] It provides an intense, sharp review of the meaning of one's life in terms of the psychological development of bonds and affinities with others. Second, the review is an important way of assessing one's social worth.[37] In this sense, the perceived *size* of the loss to family, fellow workers, and other associates allows one to evaluate the importance of oneself to those networks. But instead of finalizing this process of review by the appearance of death itself, or in the case of clinical NDE by moving further into the experience, the status passages involve *reversal.*

Reversal, the third social feature of the NDE, appears with resuscitation or rescue. It restores the person to his or her usual company and society. Attempts are then made to reintegrate this person into the everyday world of the society by rationalization and explanation of the experience. For a person with terminal cancer, this explanation might be "remission." In the case of the castaways, it may be a theory of accident, weather, or sea current patterns. The explanation offered to the experiencer of clinical NDE is often stated in terms of physiological or psychological processes. The problem with all of these attempts is that neither explanation nor rationalization approximates the *social reality as experienced.* Rather, these explanations are appropriate to the outsider's view of the experience. For example, the explanation of cardiac arrest or shipwreck is a simplistically slim description of the social reality from the point of view of those who went through this experience.

Furthermore, such rationalizations do not legitimate the social realities, and hence the experiences of the person, since these are barely acknowledged in the first place. Privation, life review, sensed presences, and the whole gamut of psychological and social experiences of dying are not the common lot of most people. Most people, then, are unaware of the range and type of social and psychological experiences that are usual or normal to this type of passage. Recounted experiences in this context commonly assume the appearance of oddity. For the experient this can lead to either stigmatized or elevated status. The realities that form a passage to another identity or value system do so because the usual social regulation of life has failed them in some way. This sense of alienation is not alleviated by explanations that do not embrace substantial parts of the experient's new social realities. As mentioned earlier, if the recounted experiences are individualized rather than seen in the context of similar, albeit marginal, social experiences, the individual's status, in the view of others, is changed.

The stories of castaways may seem heroic, miraculous, and adventurous, while those of clinical NDE may appear delusory, temporarily aberrant, and farfetched. Neither set of responses will necessarily reflect a genuine understanding of the social realities of clinical NDErs or castaways. Their status within their social networks, however, will have altered inexorably. The journey that created the status passage has propelled them from the social order of the ordinary into that of the extraordinary. From this marginal position, people who experience this kind of status passage are able to view social conformity from a new, fresh position. This is an observation made earlier by Noyes[38] and, before him, by Tolstoy in his novel *The Death of Ivan Illich*.[39]

This marginal position places NDErs in a social dilemma. On the one hand, they may be critical of the social response they receive. On the other hand, they may empathize, as never before, with the conformist, narrow view of social life that others exhibit and that was a feature of their own lives prior to their NDE. This empathy, exhibited in a variety of ways depending on different social types and responses to the NDE, may be expressed as tolerance and charity toward what they may now see as a normal, all too human limitation. The values of service and knowledge may focus on spiritual matters for clinical NDErs or survival skills for castaways; these values may contribute greatly to networks when they rejoin them. The humanitarian attitudes and value changes are those concomitant with empathy for and identification with those whose lives conform unthinkingly to the social regulation of their society.

In these sociological terms, clinical NDEs may be seen as part of a wider social experience that every member of society shares—the status passage. However, the status passage of the NDE, both social and clinical, is not normative; it is unlike transitions to, for example, parenthood or adulthood. The NDE status passage is an exception to the rule of social regulation. It is therefore a critical, or potentially critical, influence for social change. These survivors may push for greater safety and search procedures or for more support and informed understanding of life-threatening experiences. Tactical explanations and discussions of shipwreck are no more comforting to a castaway than medical ones are for clinical NDErs. Both of these are the social experiences of identity transitions; they are status passages. This means that their marginal status must be overcome through shared social understanding and historical comparison with other social NDEs. In this way, all NDEs are legitimated and their criticism is acknowledged as throwing light on the outer edges of cultural certainty and control.

Some Possible Objections

Several works compare clinical NDErs with those who experience a brush with death but without clinical NDE.[40] A review by Flynn concludes that many of those who come close to death but do not experience clinical NDE nevertheless display similar aftereffects.[41] There are frequent reports of less fear of death and a more positive and humanitarian attitude. However, Flynn observes that "non-NDErs' transformations are apparently neither uniform nor as profound as those of NDErs."[42] Flynn further suggests that non-NDErs may react to their misfortunes in ways divergent from the clinical NDE patterns. In this way, Flynn emphasizes the differences between clinical NDE and what I have called social NDE. Do the differences between the two types of NDE, which are said to be due to the variation and depth of personal impact, permit the clinical NDE to be viewed apart from a theory of status passage? I do not believe so.

The picture of clinical NDEs is itself complex. Negative clinical NDEs, for example, produce reactions (such as increased fear of death) that themselves are divergent from the usual pattern. Our knowledge of yet other differences within this variant is heavily constrained by the very small number of respondents who have been willing to discuss these negative experiences.[43] Greyson has documented that the positive clinical NDE has at least three main variants.[44]

However, these differences within clinical NDE, and between this and other social NDEs, are not objections to grouping these types together as social NDEs. Rather, attention to differences helps to define more precisely the type of NDE within an overall theory of status passage. For example, it is clear that those who experience positive clinical NDEs, and miners who are trapped in collapsed mines, tend to review their lives in past terms.[45] Castaways, on the other hand, tend to focus both on what might have been and what might still come to pass if they are rescued. The review frequently sweeps past *and* future aspects of their lives. These differences may be due to the relative prospects of rescue. Castaways must rely on meager food and water rations—with the hope, however, of renewal from sea and rain. Trapped miners, by contrast, are frequently dependent on nonrenewable and nonrationable resources such as air. These constraints provide different ways of assessing the prospects of death and rescue, and hence different ways of thinking and behaving. Negative social attitudes may also be associated with the aftereffects of some types of social NDE but not with other kinds. The clinical NDE may be an exception. The sociological reasons for this variation will be important to discover. Current differences due to variation in personal impact suggest diversity while preserving similarity. If future differences provide major challenges to similarity, then clinical NDE as a status passage should be reviewed. The present phenomenology of clinical NDEs does not appear to suggest this.

There is one final technical objection to grouping clinical NDEs with social ones. Since much of what occurs in clinical NDE seems to happen in the mind, how can such experiences be termed social in the interactive sense? First, clinical NDErs claim that despite evidence to the contrary, consciousness remains during the period in which they appear to be clinically dead. On this basis, we know that a sense of separation is felt and observed by them. Second, after their OBE, one of several interactional experiences may seem to occur. They will meet either a bright being of light or deceased relatives/friends and, with or without their aid, will review their past lives from the viewpoint of an outsider. Interestingly, some clinical NDErs claim to experience the emotions of others seen in their own life review.[46] Finally, they will rejoin their social network after the illness. In this context, they may relate their experiences and receive social reactions. In these ways, the perceived social experiences shape the emotional life of the NDEr. These experiences, as we have seen in examining the castaway experience, are not confined to clinical NDE, however unique its intensity.

conclusion

C. Wright Mills suggested that one task of sociology is to identify the types of men and women who will prevail in a society.[47] If this is so, then as a theory of context "status passage" performs this task for our understanding of some of those exposed to life-threatening experience. We need not delve into the psychology of altered states of consciousness to understand the social and cultural significance of near-death experience. Seen as status passage, near-death experiences become social experiences at the peripheries of society which are responsible for some kinds of identity transformation. Status passage as a sociological theory demedicalizes the near-death experience by emphasizing the importance of social context and comparative social analysis. Furthermore, this connecting of clinical NDE with other social NDEs, and the linking of context with reaction, allows us to see how the social shapes the psychological experience of the dying. These are the kinds of ways status passage contributes to our ongoing analysis of the NDE. But the present discussion has also demonstrated that NDE, both clinical and social, has something to contribute to our understanding of status passage. Status passages may not always be normative social structures. Rather they may also take their charges to unknown or uncharted social territories and then return them. From this experience of nearly dying, and this new status within the community, NDErs may supply new perspectives on the commonplace and banal workings of the social order. Non-normative status passages may be sources of criticism and change.

THE COMMUNITY REACTION

From Fear to Eternity

4

COMMUNITY REACTIONS

Whether from shipwreck or medical emergency, how do people react to the NDE? A review of the research on this question suggests that responses vary considerably, depending on whether one asks NDErs, their friends, or their local cleric. But the answer may also depend on when and where one asks the question. The reaction of NDErs and their social circle to this experience in the 1940s or 1960s might have been different from attitudes today. Furthermore, the reactions of contemporary Americans might be quite different from those of their counterparts in China or the Fijian Islands.

Nevertheless, it is important to understand the broad direction and pattern of community reaction to the NDE. Knowledge of people's reactions can tell us a great deal about their values and attitudes. More particularly, the NDE challenges a whole range of social relations and values, from death and God to work and gender roles. Some of this can be seen in the personal, close-up world of friends and family. Wider issues are suggested in the reaction of professionals or the media. Such issues strike at the heart of our culture's most treasured ideas about life and death.

In this chapter, I will demonstrate the diversity of these community reaction. I will attempt to show that, contrary to the popular and somewhat
dated view of NDErs as victims of negative attitudes and responses, the
NDE and the person reporting them are more often received positively,
albeit with great curiosity. For some time now, newspapers, television shows,
and Hollywood have explored, even celebrated, this popular assortment of
positive images near death.

I will begin my review with NDEr's reaction to their own experience and
then discuss the reaction of their social circle. I will summarize the little we
know about public attitudes from the survey literature and the patterns of
media production and consumption on this topic. I will conclude by summarizing the general character of much of this reaction as a preliminary to
accounting for them in sociological terms in the two chapters that follow.

Reactions of Near-Death Experiencers

We must first consider the reactions of the NDErs themselves. This is because their *interpretation* of the experience and their *changed personal values
and behavior* are the main social changes to which other people are reacting.
To simply report the sensation of being consciously outside one's body, or
to report seeing dead relatives during a period of illness or isolation, is not
a unique event. At work, for example, everyone has a story to tell. The
academic literature is replete with competing theories about these intriguing
experiences. The central problem for NDErs in a materialist world is that
many of these theories do not confirm their own interpretations of their
experience. This is also a problem with theories generated by religious people. Most clinical NDErs are convinced that their experiences are a glimpse
of life after death—at the very least, for social NDErs, an experience that
underlines the supernatural forces or possibilities in the world.

> My attitude toward death is that death is not dying; death is being reborn.
> You're reborn to a new peaceful life that when you die you will be able to
> experience.[1]

> I would say—and not being religious at all—that there must be something
> after death, which I never believed in before. I always believed that when you
> were dead, they put you in the ground and you stayed there. But I'm not too
> sure about that anymore.[2]

NDErs do not view their experiences as part of hallucination, or dreams, or as a fragment of some temporary mentally aberrant state. Nor do they believe that the devil has been playing tricks on them. Rather, NDErs believe that their experiences are spiritual experiences which reveal both God and human potential in a new and luminous light.

Sutherland, in her study of fifty Australian NDErs, reports that the doctrinal aspects of organized religion no longer have any appeal. The ideas of angels with wings, strict church attendance, and divine judgment appear medieval and childish.

> They say if you're not a Christian none of you will be able to come in through the eye of the needle, and all that sort of thing. And I think, well, I went up there and I saw it and I certainly wasn't a Christian at the time. So how do they know?[3]

In a letter sent to the *Journal of Near-Death Studies* an NDEr remarks:

> . . . the NDE is not a Love or God experience. It is rather an event that allows us to experience pure Being, a state in which judgment has absolutely no place.[4]

The belief in life after death, for NDErs, is associated with broader beliefs about life: the positive and loving nature of God; the personal worth and destiny of every individual; the possibility of reincarnation; and the reality of paranormal abilities such as extrasensory perception (ESP). Atwater, an NDEr who wrote about her own experience near death and that of others, reported the following:

> Some of the more unusual phenomena involve survivors who claim they now regularly meet with the light beings or angels they saw during their near-death experience. Others claim to see and talk with departed spirits of the dead. Many claim to see plants and especially flowers undulate as if breathing, while still others claim to see a web-like substance connecting all in sight with everything else through a network of glistening threads.[5]

More commonly, however, the psychic phenomena are less spectacular and more mundane. Flynn quotes one NDEr on this subject:

> Most of these things have a matter-of-fact quality about them. I just know they are going to happen and don't bother to ask questions about them. Lots of times, it's just little things. I'll pick up the phone before it even rings and there will be somewhere there.[6]

To NDErs, the NDE is a spiritual, perhaps even mystical, experience that allows them to consider or reconsider the material world in ways that include paranormal possibilities in themselves. This view of their NDE is the fundamental stimulus for change in personal values and behaviors. NDErs give material concerns lower priority. The active pursuit of status or money is often abandoned.

> Success to me isn't spelt out in material terms even though over the years I've done very well materially. It's spelt out to me in terms of happiness. I could be a poor man writing verse on the beach and still be very happy. I think it comes down to the fact that I just love the pleasure of being here. Just the fact that I can see magnificent scenery and talk to good friends and enjoy good company is great.[7]

> My interest in material wealth and greed for possessions was replaced by a thirst for spiritual understanding and a passionate desire to see world conditions improve.[8]

Concern for others increases, often prompting some NDErs to give generously of their time to causes and persons previously strange to them.

> The most important lesson is that you stop and help people (all living things) by encouraging and comforting, trying to understand why people act mean and cruel. Then, you help them see that they are doing and justifying all acts from fear. It's better to do everything out of love (although I still don't always love all the time myself . . .).[9]

> I now breed day lilies for fun and give away hundreds yearly.[10]

Respect for and attraction to all forms of life and knowledge develops, and learning becomes an active pursuit. This interest appears informally, in the form of greater attending and listening to others, and formally in terms of interest in college or other institutional courses of instruction.

> I don't shoot deer, or hurt animals, or cut down trees anymore, because I can feel for them.[11]

> My purpose and my outlook on life became from that point on a searching . . . books became my friends (before this, she said, she tended to read only escapist literature) . . . I found myself on a college campus, which was somewhere that I always wanted to be when I was younger, but I never got to do . . . I went back to school. . . . [12]

Knowledge is important. I read everything I can get my hands on now, I really do. It's not that I regret taking the path I did in life, but I'm glad that I have time now for learning. History, science, literature. I'm interested in it all.[13]

Fear of death all but evaporates. When sociologist Cherie Sutherland asked her respondents about this, "Many laughed at the question."[14] Some of Raymond Moody's interviewees spoke of it this way:

I was only a child when it happened, only ten, but now, my entire life through, I am thoroughly convinced that there is life after death, without a shadow of a doubt, and I am not afraid to die. I am not. . . . I've had many things happen to me in my life. In business, I've had a gun pulled on me and put to my temple. And it didn't frighten me very much, because I thought, "Well, if I really die, if they really kill me, I know I'll still live somewhere."[15]

Now, I am not afraid to die. It's not that I have a death wish, or want to die right now. I don't want to be living over there on the other side now, because I'm supposed to be living here. The reason I'm not afraid to die, though, is that I know where I'm going when I leave here, because I've been there before.[16]

How do friends and family respond to an aggressive, competitive, and materialist man who suddenly talks about love and service to others? How do they react to his recent interest in growing and giving away day lilies? How do friends and work colleagues respond to their life-of-the-party swinger who, since her recent appendectomy, has suddenly become introspective, vegetarian, and psychically sensitive? Understandably, their responses and adjustments can be difficult. But are they mainly negative?

Reactions of Family and Friends

Raymond Moody's landmark book on the subject of NDEs gives the impression that the response to NDErs' accounts of their experiences was mainly negative.

You learn very quickly that people don't take to this as easily as you would like them to. You simply don't jump up on a little soapbox and go around telling everyone these things.[17]

However, closer inspection of Moody's report reveals a more complex picture. First, it is not clear whether the negative responses were to the simple description of the NDE or to the experient's interpretation of it. As mentioned earlier, it is one thing to tell people about the shapes, events, or feelings in a dream or an NDE. But if an unusual interpretation accompanies the description, one can expect to raise some eyebrows. If claims are made about *actually* traveling to a world beyond the grave or *actually* meeting Jesus, the storyteller can certainly expect surprise at best, resistance and stigma at worst. In this case, the question of what it is that people "don't take to . . . easily" is an open one.

Second, it is not clear from Moody's work how representative or widespread the initial responses to the NDErs were.[18] This is because several of them simply stopped relating their story when early reactions were unsatisfactory to them. In several cases this may have been unwise. For example, of the eight NDErs who received negative responses:

One was a child whose story gained little attention from his mother.
One had a minister who stated that the experiences were probably hallucinations.
One told her story to fellow school children in "junior high and high school."
One had nurses who believed that these experiences were imagined.

Few of these persons could reasonably expect a sympathetic hearing, with the possible exception of the mother of the young child. Even here, however, we do not know the circumstances that led to the apparent lack of interest. School children are not known for sensitivity; they wish to conform to peers and fear rejection. Ministers do not take kindly to competing ideas of the afterlife or, depending on the denomination, to the very idea of survival of death. And health professionals tend to categorize unusual psychological experiences in terms of theories of psychology.

These reactions from family, friends, and some professionals were unfortunate because they encouraged some NDErs to remain silent, not giving others, who were possibly more interested and sympathetic, the opportunity to hear and react positively. Two further NDErs who made up Moody's eight cases said nothing to anyone for years, simply out of *anticipated* rejection. But to conclude purely from these cases, and others like them, that the community reaction to NDEs and NDErs is poor or unreceptive is premature. We can gain greater insight into these matters from Insinger's sociological study of the impact of NDEs on family relationships.[19]

Insinger interviewed eleven NDErs about the impact of their NDE on family members. Once again, the results are not generalizable; that is, we cannot say what proportion of the general population might share these reactions. However, the study did demonstrate a more diverse set of reactions; these were either positive or neither positive or negative.

> One NDEr's family accepted the reality of the NDEr's experience but not necessarily the interpretation.
> Three NDErs did not tell certain family members because their relationship was poor in general.
> One NDEr told her mother, who did discuss the experience with her but then did not raise the subject again. This was interpreted negatively by the respondent.

These were often adult children referring to their parents' reaction. Some of these relationships were not optimal. However, Insinger points out that these same NDErs often received positive responses and support from their spouses or children.

> I've had no problems with my family at all. I don't know that they said, "I believe you," but they never said they didn't—and they accepted it—whether they believed it or not, they accepted it.[20]

Three NDErs became divorced or separated because, according to them, the NDE had changed their values and outlook on life so much that their spouses could not reconcile themselves to these changes. But again, a closer inspection of the interview material reveals a more complex picture. The first divorcee revealed that

> she thinks they were probably destined to be divorced eventually because of his psychiatric problems, but that she remained in the marriage for a while, hoping that things could be worked out.[21]

Another NDEr who separated from her first partner reflected that

> we had a sort of iffy relationship; we didn't know if we were going to get married or not, but after my experience, we absolutely could not relate to each other in any way, shape or form.[22]

The last person interviewed remarked that her former spouse was used to a dependent wife, a wife who, after her NDE, no longer related to him this way.[23]

We must separate negative responses to the NDE from negative responses to changes in a poor relationship. In the latter case, the death of a child, a lottery win, or a change in the educational level of one partner can trigger the process of separation. In that context, the NDE is one of the many exigencies of life which may bring people to alter their relationship, accommodate to the change, or be swept apart by it. It is not that a particular characteristic of the NDE is troublesome in these cases, but rather the mere presence of some (almost any) significant change in the relationship.

Insinger's insightful work reveals some of the possible impacts of crisis and personal change on a relationship. Friends and family may

React negatively or positively to the NDE account.
React negatively or positively to the NDEr's interpretation of the account.
React in a neutral or uninterested way.

Important in predicting the possible direction of this reaction is the state of the relationship. Both Moody's and Insinger's accounts seem to suggest that the NDE will

Exacerbate an already poor relationship or
Improve an already effective and supportive relationship.

Still, these few cases and anecdotes do not give an impression of the *pattern* of community reactions. We know that these reactions will vary, but we do not know which, if any, of these will predominate. Only by understanding the pattern of community responses can we attempt to unravel what may appear to be idiosyncratic and random. There may be reasons for people's reaction to NDEs that originate in our cultural institutions and organizations. This is where surveys play a particularly useful role.

Reactions of the Community

Because no one had ever assessed public attitudes to NDEs and NDErs, a colleague and I decided to conduct a small survey.[24] We designed a questionnaire that began with a short vignette of the popular idea of an NDE. This vignette described an accident or medical emergency in which an individual recounts the five main elements of an NDE: tunnel sensation, an OBE, meeting deceased relatives, meeting a bright being of light, and experiencing a life review. The questionnaires were divided into two types: those that portrayed "a close friend or family member" and those that portrayed "an acquaintance."

The next section of the survey contained seven possible explanations of these experiences. The final section asked people to choose from a list of possible reactions, such as "I would humor this person until the delusion passed," "I would encourage discussion of his/her feelings, whatever these might be," or "my reaction would depend on this person's reaction." The survey was distributed to 174 members of the general community.

Analysis revealed an even spread of age and educational level, with a fairly even number of male and female participants. Some of the results are summarized in Table 4-1. Note that if life after death is a desirable interpretation for the NDEr, the result reported indicated plenty of community support. Interpretations that might be considered negative by NDErs—hallucinations, mental illness, or just imaginings—comprised only 14 percent of the total response.

Also interesting was the fact that there was no significant difference between the "acquaintance" and the "close friend or family member" surveys. People appeared to explain an NDE or to respond to someone who had experienced one in a similar way irrespective of their relationship. Scores on the reaction part of the survey revealed that most responses to an NDEr would be positive. This positive reaction was associated with younger people in general, female respondents, and those who believe in life after death. Conversely, older people, men, and those who did not believe in life after death were more likely to react negatively, that is, in rejecting or avoiding ways.

If this small sample of reactions is any indication, NDErs should come across significantly more positive reactions than negative ones. This finding, however, does not appear strongly in the anecdotal discussions on this topic. Why not?

First, many people who, like those who choose what might be seen as negative explanations, do not know how to explain the NDE. These people make up some 24 percent of the sample. This is a proportion high enough for any NDEr to encounter regularly. Two possibilities then emerge. As I have stated in an earlier study of terminal illness,[25] negative reactions are traumatic, strongly influencing personal recollections. In other words, if associated with trauma, negative images can dominate an otherwise diverse and largely positive range of experiences associated with an event. An alternative or additional factor may be that, as mentioned earlier, early rejection

Table 4-1   Responses to Seven Explanations for the NDE

It was a passing hallucination.	9%
It was a dream.	6%
It was the beginning of a mental illness.	2%
It was the side effect of medical drugs/techniques.	4%
It was possible evidence of life after death.	58%
It was the product of a vivid imagination.	3%
I don't know how to explain it.	18%

may preclude some NDErs from sharing their experiences with others who might have been more positive.

Second, the majority of the survey group believed in life after death. This was important to their attitude that the NDE was an indication of this possibility. This large proportion of believers is fairly typical. The pollster George Gallup conducted a survey of over 1,500 households and found that 67 percent of people believed in life after death.[26] This figure is even higher for churchgoing people. At least a quarter of Gallup's sample also believed in reincarnation and communication with the dead. Many of these respondents reflect what has sometimes been called *New Age* attitudes. New Age beliefs include beliefs that differ significantly from both modern materialism and local religious orthodoxy: a belief in Buddhism, reincarnation, astrology, spiritualism, and the music, literature, and lifestyle habits that flow from these beliefs.[27] Presumably, these people generally also believe the NDE to be an indication of life after death.

There is a small and unpopular group in many Western societies who do not share other people's paranormal or occult explanations for these experiences, preferring instead to explain them in psychological or physiological terms. These people sometimes band together in belligerent organizations that discourage uncritical belief in these issues. The members of these organizations are known as *skeptics*. Although they are very vocal, they are numerically few compared to believers.[28]

But if skepticism characterizes only one person in four of the general community, the odds favor the majority of NDErs enjoying positive responses to their accounts. The fact that this does not seem to happen might suggest that the one-in-four figure does not occur randomly in the community. Views of the NDE that do not embrace the possibility of life after death may occur more frequently in some groups than in others. It may be possible, then, that NDErs are exposed to this group or groups soon after the NDE. I am thinking, of course, of two groups in particular: doctors and older people (such as parents).

Some evidence for this suggestion comes from Gallop's survey of doctors; some 68 percent of them indicated that they did not believe in life after death, and a further 16 percent held "no opinion."[29] This is almost the reverse of the survey findings of the general public. Education does not appear to be positively or negatively associated with different interpretations of the NDE, but clearly the beliefs of those with a medical education preclude the possibility of life after death. NDErs are much more likely to receive negative interpretations from this group, and of course, they often encounter doctors soon after a clinical NDE. However, medical people are not the main or even the only group who will respond this way.

My survey also indicated that some 79 percent of respondents were familiar with NDEs from published stories, and many of them had also encountered the subject on television, film, or the radio. Once again, familiarity with and a positive presentation of this subject do combat ignorance and fear, whatever else one might say about the biased nature of the programming. I will discuss this further later in the chapter.

Finally, of course, and related to the issue of familiarity, is the age factor mentioned earlier. Younger people are great television watchers. They are also better educated than older ones. The positive reaction of one daughter of an NDEr is attributed by her to familiarity with NDEs from a college course:

> I think the reason that I did was just having had that course. If I hadn't had that course, I think I would have been in a different spot—because it would have been such a foreign idea to me—but having had that and realizing that other people had it and all that stuff, it was kind of exciting.[30]

Older people, born before the advent of television, may be less familiar with the medium's obsession with the NDE and the occult. In any case, certainly before the current explosion of interest in death and dying that began in the early 1970s, many people, perhaps most, may never have read or heard of the things reported by today's NDErs. It is difficult to assess the role of information here, but presumably regular reports of the experience might "normalize" one's attitude by encouraging the view that such events are common and harmless. In that context, NDErs who related their stories before the post–World War II rise of television might have been far more likely to attract surprise, awkwardness, fear, or ridicule. This reaction is less likely today. However, media information notwithstanding, older people tend to be more conservative in their views than young people in surveys on almost any issue. There are echoes of these influences of age and media in another survey conducted by my colleagues and I.

Table 4–2   Responses to Eight Explanations for the NDE

It was a passing hallucination.	39%
It was a dream.	19%
It was the beginning of a mental illness.	5%
It was the side effects of medical drugs/techniques.	4.5%
It was possible evidence of life after death.	9%
It was the product of a vivid imagination.	9%
I don't know how to explain it.	12%
Other possibilities (please describe).	2.5%

We repeated the above survey in China the following year in an attempt to gain some comparative insight.[31] A total of 197 people in and around Beijing participated, and the pattern of preference for explanations was the reverse of the Western one. I summarize the results in Table 4-2. Note here the preference for materialist explanations and the unpopularity of life after death as a preferred explanation. Once again, older people tended to respond negatively. There are several possible reasons why the Chinese take a more conservative approach to the subject of NDEs. Lack of media exposure, especially of the occult-oriented type commonly associated with NDE coverage in the West, may be an important factor. Only 24–31 percent of Chinese respondents had heard, seen, or read about the NDE, a remarkably low figure by Western standards. Nevertheless, this may not be the only, or even the most important, influence on these attitudes.

Cultural differences related to Chinese conceptions of the afterlife may be more important. Also, because surveys are usually conducted by government officials rather than by commercial pollsters or academics, the respondents may simply have wished to appear ideologically sound. Materialism is the state ideology, and religion and superstition are discouraged. The responses to our questionnaire may have been politically correct, but were they a fair indication of actual attitudes? Given the primitive nature of our survey, we are not likely to settle this question without more sophisticated methodology.

However, returning to the question of general public attitudes, particularly in the West, most people appear to be interested in the NDE. Furthermore, most people seem to concur with the NDEr's explanation. Apparently, however, certain segments of the community deviate from this position. Unfortunately for the NDEr, the people they often attempt to tell first fall into this second category. These are often people who are likely to be less informed and more conservative in general (pre-baby-boomer parents), or people who are atypical in their attitude toward issues about death (doctors), or people who have a vested interest in maintaining specific images of the afterlife (clergy). Once again, to characterize "society's" reaction to NDEs or NDErs based on the responses of these people would be to overgeneralize from atypical sections of the general public. I now want to take a closer look at some of these sections of the community, beginning with the clergy and ending with a discussion of the media.

The clergy

The religious tradition that has dominated the cultures of the industrial West for the last few thousand years has been the Judeo-Christian one. That tradition, as the religious scholar Carol Zaleski notes,[32] has been regularly punctuated with stories of "otherworld journeys" ranging from those told by St. Paul and Gregory the Great to later medieval ones such as Dante's adventures and those of Drythelm the monk of Wenlack. Some of these stories have been interpreted as evidential to the Christian church's moral teachings, their life-after-death detail serving as the medium of instruction. Others have interpreted these accounts as testimony to the faith and witnessing power of the saints and the power and glory of God. But usually those stories, drawn from visions inspired by ascetic practice or possible NDEs, are viewed positively.

Even today, religious literature from the main Christian churches seems to be responsive in ways that, if not reassuring to the NDErs, are at least not in direct conflict with their strongest feelings. The outspoken Catholic theologian Hans Kung has written that the clinical NDE is not necessarily evidence of life after death because these are not matters decidable by evidence but rather by faith.[33] NDErs did not die because death means the inability to return. But Kung does welcome the NDE for another reason:

> The positive experiences of dying create a hope that dying—normally awaited with apprehension, even with fear and trembling—in its very last stage may possibly not be as fearful as is often anticipated. Perhaps the changes in the features after a severe death struggle, which so often make the countenance of the person now really—biologically and not merely clinically—dead seem so peaceful, relaxed, even smiling, "blissful," is a sign—but no more than a sign—that a new existence is not a priori to be excluded: a sign of a transcendence of death.[34]

The Presbyterian theologian John Hick, although not writing about the NDE specifically, does take seriously the contribution of both spiritualism and parapsychology, including crisis apparitions, in his assessment of life after death.[35] In his famous book *The Power of Positive Thinking*, Norman Vincent Peale devoted a whole chapter, entitled "Prescription for Heartache," to crisis apparitions and deathbed apparitions.[36] Clearly, both of these religious leaders would have integrated the NDE in similar ways into their treatment of these related experiences. And Billy Graham, probably the most famous of the Baptist evangelists, has given his listeners legitimate interest in NDEs:

> Dr Raymond Moody has interviewed more than one hundred people who have medically died and come back to life. He has put these interviews in a bestseller entitled *Life After Life*, which comes the closest to illustrating what it is like to die. Still, it is not an easy thing for people to describe.[37]

The pronouncements of theologians and celebrity preachers do not always indicate the general clergy's reaction to NDErs. There have been two major surveys of the clergy's attitudes toward the NDE. David Royse surveyed 174 clergy, over half of them Methodist but also including other denominations such as Catholics.[38] He found that some 71 percent of them had been approached by parishioners to discuss NDEs. About three-quarters of his sample did not feel that it was a problem to introduce the topic of NDEs to dying people, nor did they believe that NDEs conflicted with biblical teachings. The majority of respondents felt that NDEs were an actual glimpse of the afterlife. This group would surely react positively to the experiences of NDErs.

More recently Lori Bechtel and some of her colleagues surveyed 320 clergy about their attitudes toward the NDE; they also found that this attitude was generally positive.[39] However, the English archdeacon Michael Perry objected to that conclusion because the survey was based on a 12 percent response rate.[40] Perry seemed less optimistic about the clergy's attitudes to the NDE. He may have been reacting, once again, to the powerful early anecdotes discussed previously and/or perhaps to the objections from some Christian fundamentalists.

Some religions, such as Seventh Day Adventism and some strands of Judaism, do not believe in survival after death. Adventists, for example, claim that survival must wait for the final resurrection of the body at the end of the world, during Christ's second coming. Other religions do not regard the NDE as evidence of survival at all. They consider NDErs' interpretation as somewhat contrary to biblical teaching and, instead, side with some medical explanations. The Worldwide Church of God, well known for its publication *Plain Truth*, tell it the following way:

> Though many of these happenings are understandably astounding and seem to contradict what the Bible says about death, the whole idea of this so-called "life after life" experience is based on the premise that these people died.
>
> According to the medical profession, some of these individuals were clinically dead. However, medical science has not yet agreed on what constitutes actual death in a human being. Real death, according to the Bible, is total unconsciousness — without memory, feeling, knowledge or perception (Ecclesiastes 9:5, 10; Psalm 6:5). It appears, therefore, that those who were revived to relate their experiences may not really have been dead, but simply in an unconscious state.[41]

Gallup, in his review of different religious reactions to the NDE, also mentions a religious interpretation that emphasizes the trickster nature of Satan. All the major features of the NDE "fit perfectly with Satan's clear purpose to counterfeit the truth of God's word."[42] The Christian writer Kerby Anderson, in his own examination of the near-death studies literature, concurs:

> It is interesting that in turning to 2 Corinthians 11:14–15 we read that "Satan himself masquerades as an angel of light" while "his servants masquerade as servants of righteousness." The being of light and the other people seen in these paranormal experiences could indeed be Satan and his fellow demons who are working to perpetrate a spiritual hoax on those who are near death and on the living who hear these reports.[43]

These interpretations of the NDE, then, are designed to encourage the NDEr to feel tricked either by Satan or by his or her own brain or psyche. So again, like the general public, the clergy seem generally positive to the NDE, but they offer conflicting interpretations. These can give rise to significant stigma and disappointment for the NDEr. The chances of encountering an interpretation congruent with the NDEr's however, are nevertheless far from random. Specific religions, for example Adventism, and specific traditions, for example some types of Christian fundamentalism, are more likely than others to hold views of the NDE that contradict the NDEr's ideas.

The Health Professions

We have already touched on the issue of medical attitudes that do not share the NDEr's interpretation of the experience. But beyond this, some medical theories appear to infer weakness or poor psychological mastery or control. Gallup[44] provides some examples of these professional attitudes:

> . . . [T]he experiences are *psychoneurotic,* imagined or expected on the basis of belief.
> They are very sincere, but usually the descriptions are *plagiarized* or borrowed from others.
> The ideas of lights, ghostly figures, religious symbols, etc., are firmly implanted into the minds of most *children* at an early age, and perhaps the brain releases them under such stress.
> (Emphasis mine)

Is it any wonder that some NDErs are taken aback when medical opinions such as these suggest that the NDEr is either neurotic, self deceived, or regressing to childlike fantasies? Michael Sabom, a cardiologist and author of *Recollections of Death: A Medical Investigation,* admitted that his earlier attitude prior to his study was disbelief even of the existence of the experience.[45] But medicine is not only a profession (and a mainly male one at that); it is also a science. As a science, it is imbued with the materialist assumptions of the day, and these values encourage questioning of nonscientific theories, that is, theories other than the ones they endorse. We will examine this more closely toward the end of this book when we discuss the academic response to the NDE.

The nursing profession is mainly female, and its allegiance to scientific models of inquiry is not so unquestioning as that of medicine. This might suggest a more diverse and possibly a more sympathetic view of both the NDE and the experient's interpretation of it. This does seem to be the case. The nurse researcher Annalee Oakes conducted a small survey of critical care and emergency nurses in the mid 1970s.[46] She did find strong negative reactions to NDErs, reflected in such comments as "preposterous," "a scam," "religious nuts," and "psychiatric blow-outs." But the overwhelming majority of respondents were fascinated and desirous of learning more about the experience. This early work is supported by the larger survey work of another nurse researcher, Roberta Orne.[47] Orne surveyed 1,600 nurses and drew a response from over 900. Most of them were familiar with the NDE, and most exhibited positive attitudes to the NDE and NDEr. Most of those surveyed would "encourage discussion" of their experiences (86 percent) and offer support (69 percent).

Clearly again, there is a small core of people with rejecting attitudes. I am referring not to those who simply do not share the NDErs' interpretation of their experience but to those who, on hearing of the experience, immediately view the NDEr as socially or psychologically incompetent. Those who did not respond to these surveys may also disguise the approximate size of this stigmatizing group. The pattern of nursing attitudes to the NDE seems similar to that of the general public: generally positive, even openly fascinated, but with views markedly different from this held by a sizable minority within their ranks.

Finally, Barbara Walker and Robert Russell surveyed clinical psychologists about their views of the NDE.[48] From over 300 mail surveys, these researchers received returns from 117 respondents. This survey showed that, a few antagonistic responses aside, clinical psychologists appear overall to exhibit a positive attitude to NDEs. The pattern here appears similar to those revealed in surveys of nurses and the clergy but different from the results from medical respondents.

In summary, then, within the general community, older people, men, and those who profess not to believe in life after death are more likely to exhibit negative attitudes toward the NDEr. Among members of religious groups, the general view of the NDE is similar to that of NDErs themselves, with the exception of some religions, where this contradicts their stated eschatology. Mostly this applies to certain fundamentalist groups. Among health professionals, the attitudes reflect the diversity and general positiveness of the rest of the community, with the notable exception of the medical profession.

Behind most these survey results, whether of the general community or of professionals and clergy, the main source of knowledge about the NDE appears to be the print and audiovisual media. Surprisingly few people report having read even one of the main books on the topic—those by Moody or Ring, for instance. Even fewer, of course, have examined the journal literature, either thanatological or medical (the medical attention to this literature will be commented on later). Aside from knowledge gained by meeting an NDEr, much of the principal information seems to be derived from popular magazines or newspapers. At other times, a television or radio program will be the source. What, then, is the main "angle" adopted by the media on this topic, since this source is so influential with such large, diverse groups?

The Media

The reaction of the media to the NDE has been nothing short of spectacular. Few would disagree with the observation that not a week goes by in America without a television, radio, magazine, or newspaper story on this experience. Furthermore, the film industry now regularly produces, for the big screen or video market, movies that depict images of the NDE as an integral part of their story.

Flatliners, the story of medical students stopping and restarting each other's hearts to explore the experience of death, is probably the most well known in this genre. The highly successful Australian film *Bliss* shows how an NDE can change a person's life. *Resurrection* is a film about the NDE and the subsequent life of one woman who develops psychic powers from the experience. More recently, *Defending Your Life* is a film that integrates some imaginative afterlife imagery, with the life review being central to the plot and outcome. Philosopher Michael Grosso mentions still other films that use some or many parts of the NDE. Among these are *Return of the Jedi, All That Jazz, Peggy Sue Got Married, Jo Jo Dancer, Somewhere Tomorrow, Ghost*, and *Jacob's Ladder*.[49]

It is difficult to keep track of all the NDE material in the print media (there's so much of it in different U.S. states and other Western countries), but a browse through a few of the newsletters of the International Association for Near-Death Studies gives some indication:[50] *Encyclopedia Britannica, Life* magazine, *The Saturday Evening Post, Washington Post, Palm Beach* (Florida) *Post, South Carolina Courier/News and Post*, and *The Philadelphia Inquirer*.

On television, shows covering the NDE appear on national stations from Philadelphia (*A.M. Philadelphia*) to Tokyo (Nippon and Fuji TV) to Sydney (ABC radio and TV), and this does not even take into account major U.S. shows that are exported. *Oprah Winfrey, Geraldo, Sally Jessie Raphael, The Phil Donahue Show, 20/20, Good Morning America, 60 Minutes, Hugh Downs, The Joan Rivers Show*, and *Unsolved Mysteries* are just some examples of these.

Most of these shows, films, and literature cater to the large number of readers and viewers who, as we have seen in community survey results, believe in life after death. But more than this, people are enthralled that a person's life and attitudes can be transformed for the better as a result of one single, powerful experience. And the personal transformation generates social values that, for the educated post–World War II generations of the

West, resonate strongly. Talk that generates images of a new world (here and on the "other side") based on cooperation, knowledge, and caring rather than the centralized authority of church or government, has regularly appeared in a variety of twentieth-century ideologies. It has been particularly appealing to the baby-boomer generation raised on 1960s ideas of humanism and liberalism.

In that context, TV shows and articles use skeptical theories and theorists as stirrers and party-poopers. Critical voices appear menacing; objections to parts of some explanations appear to dismiss the whole experience. TV time and newsprint space tend to favor the NDE account, giving the critical or alternative view the role of sneezing powder toward the whole affair.

An amusing but instructive case is offered by Robert Basil.[51] Mr. Basil, a former editor for Promethous Books and a member of an organization of skeptics, was invited to appear on *The Shirley Show* during its coverage of NDEs. I will let Mr. Basil relate the rest of the story:

> The hostess introduced me with this question: "Over 8 million people report having had a near-death experience. Why do *you* say they're lying?"
>
> This debate, I realized at once—with considerable force—was going to be rather limited. To doubt the prevalent interpretation of the NDE meant that one was an atheist, that one had no hope, that one's mind was closed. The near-death experience was the subject of the show only insofar as it cleared the way for discussion about spiritual growth and faith-proved-true. I *did* note, in fact, that many NDErs return from their experiences with feelings of universal love, that there was no need to attack the experience itself, especially when its after effects were so manifestly beneficial. Nonetheless, the hostess nailed me with the kind of logical *non sequitur* that plays so well on television: "Don't you believe in love, Mr Basil? Haven't you ever been in love?"
>
> The cameras were showing audience members shaking their heads, apparently with disbelief and derision, as I pathetically protested, "I *love* Love!"
>
> I'm sorry, that really wasn't fair of me," the hostess said. "So let me ask my audience: is there anybody here who agrees with what Mr Basil is saying today?" After five seconds of silence—a long, long time on television—she summed up: "Okay then . . . We'll be right back after this message."[52]

The stylistic approach seen above and the frequency of media coverage of the NDE suggest that both the experience and those who have it are in heavy demand. From an entertainment point of view, the NDEr is a celebrity in two important ways. First, the NDEr is a person who has had a highly unusual, and hence fascinating, experience: dead for 14 minutes but resuscitated; adrift at sea without food or water for 29 days; lost in the Himalayan snows for 16 days with only a chocolate bar; and so on. These are the kind of events and experiences that release us from our boring, predictable workaday world. These interviews function in the same way as an Arnold Schwartzenegger film or a Fourth of July fireworks show in New York's Central Park. These are riveting and engrossing experiences, their unique features and novel sensations capturing and refreshing our somewhat jaded and flagging attention. Second, NDE accounts help people reflect on issues of personal significance: values, attitudes, direction in life, possible sources for personal change or development, or hope, or happiness, and so on. The account of the hazardous journey, undertaken unwillingly by an ordinary person but ending in survival and return, is inspirational. Its very construction leads to listener or reader identification, providing hope in the context of the dozens of experiences so many of us endure, unwillingly, in our own life journeys every day. That is the magic and the social attraction of NDE accounts for the media. It is the NDE story itself, not its possible explanations, that attracts readers and viewers. That is one reason why progress in clinical, social, or biological sciences is not as alluring as the NDE accounts told by the NDEr themselves. We have now come full circle in our examination of the community's reaction to the NDE.

conclusion

Although the social circumstances surrounding those who experience an NDE appear rather complex, a clear direction and pattern of public response is discernible. Negative public reactions to NDEs appear to characterizes disclosures of the 1970s or earlier. These were times when our information about the NDE was rather poor. People may have had difficulty distinguishing such accounts from their own folk understandings of mental illness or ideas about the human imagination.

But it is also important to remember that the NDErs themselves were not immune to these influences and were themselves sometimes doubtful about the nature of their experience. These concerns, aside from the reactions of others, may have prevented many NDErs from sharing their experiences.

Others began to share their experiences with trusted people who might be thought to know about these things but were disappointed with the reactions received, from doctors, clergy, or parents in particular. During the 1970s and 1980s, and as the media spread the rather sensational news that

NDEs might be evidence for life after death and an impetus to spiritual growth, attitudes to the NDE took a strongly positive turn.

Current surveys of the public and many specific sections of the community reveal overwhelming support for the NDE and the NDEr. Even people who are skeptical of the life-after-death interpretation of the NDE do not resist the notion that the NDEr, like anyone who undergoes a major crisis, needs support and opportunities to discuss and share their experiences. But more than this, newspapers, popular magazines, films, TV, and radio regularly cover this topic. And their coverage is often in sympathy with the interpretation of NDErs, sometimes explicitly playing up to New Age ideas that are widely entertained and accepted. A critical, balanced, and informed discourse is not to be found in current media coverage on this topic.

However, this fanfare and popularity is itself a result of several important historical and social developments in the last twenty-five years. The popularity of the NDE is not simply about the gullibility of human nature, nor is it about the uncontrolled spread of New Age ideas. Gullibility is in the eye of the beholder, and the acceptance of New Age ideas should also be understood as a part, rather than as a cause, of the changes that make the NDE so popular. If this is so, then what are the reasons why the NDE has enjoyed so much popularity in the last twenty-five years since the appearance of Moody's *Life After Life?* We will now turn to the major social changes that I believe are responsible for this development.

5

SOME RHYME AND REASONS

London Bridge is falling down,
falling down,
   falling down…

It is tempting to think that the major reason for our fascination with the NDE has to do with what some people have described as our long-standing fear and denial of death. This is a reason eagerly embraced by an impatient, psychologically oriented public. Nevertheless, there are other reasons for supposing that more complex influences lay behind the popularity of the NDE.

Like the children's rhyme "London Bridge," things may not be what they seem. To the psychologically oriented, this rhyme may be about how children play at cheating death, singing a rhyme that harks back to a time when children were sacrificed in the hope that their souls would guard and protect bridges.[1] However, almost no child who sings this rhyme today understands or cares about its origin. The rhyme is not sung for this purpose. Its perennial popularity is not explained by the presence of this death imagery, its psychoanalytic possibilities notwithstanding.

Similarly, the popularity of the NDE cannot be explained by the mere presence of afterlife imagery. This is because both afterlife images and NDEs have existed for millennia. Why, after all this time, should NDEs gain so much popular attention? An understanding of other influences is necessary to answer that question. In this chapter and the next, we will examine five of the main sociological reasons for the current popularity of NDEs. The four reasons discussed in this chapter are technical and demographic changes underlying our experience of death, changing attitudes toward death and dying, the changing role of religion, and our changing relationship with social institutions.

Technical and Demographic Changes

One reason why NDEs are so widely talked about has to do with the large number of people who are recounting them. And since resuscitation is now such a widespread medical procedure, everyone is listening. It is estimated that some 400,000 to 600,000 deaths each year in America are due to sudden cardiac arrest.[2] The majority of these people are subject to resuscitation procedures. This is because cardiopulmonary resuscitation (CPR) is initiated on most people irrespective of the underlying disease process.[3]

The possibility of legal proceedings against anyone who does not provide a patient with every possible life support is also a stimulus to these practices. Even outside the hospital, the physician Hermreck[4] recommends CPR

> for anyone not showing reliable signs of death such as decapitation, rigor mortis, evidence of tissue decomposition, or extreme dependent lividity.

This policy and these techniques have enjoyed such currency only since the early 1960s. Although resuscitation has a long history, its success (such as it is) in reviving those whose hearts have apparently stopped beating is quite recent. The combination of techniques currently in use emerged from a rather depressing history of trial and error that continued up to the mid-1950s.

There are accounts of resuscitation in Egyptian mythology, and Hebrew midwives used expired air to resuscitate newborns.[5] There are even biblical accounts of resuscitation.[6] From ancient Egyptian to medieval European times, a variety of techniques, usually unsuccessful, were attempted. These included smoke enemas; being hung upside down by the feet; bottles of hot water; buckets of cold water; yelling and screaming; the application of animal excreta, holy oils, or amulets; tongue stretching; being placed on a trotting horse; inflation of lungs employing a fire bellows; and mouth-to-mouth exchange of air.[7]

The current system of resuscitation involves Airway control, ensuring that the airway is clear of foreign matter or anatomical structures such as the tongue; Breathing control, usually administered in the field by mouth-to-mouth techniques and in the hospital by mouth to tracheal tube or mask; and Circulation control, which usually involved closed-chest massage. These "A-B-C" steps are usually followed or accompanied by "D-E-F" procedures: Drugs and fluids; Electrocardiographic monitoring; and Fibrillation treatment involving electroshocks to the heart to restore the normal sinus rhythm.[8]

In addition to the modest technical success of current CPR methods, millions of people have undergone training in their use in order to apply them in the workplace, at school, and in public places such as restaurants and pools. Simulators, or manikins, for the practice and demonstration of resuscitation were developed and marketed by the Norwegian doll maker A.S. Laerdal. "Resusci-Anne," his first doll, was a life-sized figure of a young adult female used to teach mouth-to-mouth resuscitation.[9]

Other models followed, including Resusci-Baby and Resusci-Junior, which allowed whole generations of lay and professional populations to learn how to resuscitate one another. In medical settings, artificial torsos were manufactured for defibrillation practice, allowing health care staff to improve their accuracy in applying electric paddles to the chest.[10]

Advances in training methods and their spread among lay populations means that resuscitation is not confined to medical settings. But aside from the vastly increased number of people who can apply CPR, there is the issue of how many times it can be used. Resuscitation can be have multiple applications. President Dwight Eisenhower, for example, underwent fourteen resuscitation procedures in one week when he experienced as many episodes of ventricular fibrillation.[11] So, the large number of resuscitation cases makes NDEs a significant likelihood. However, the number of applications to a single person also makes it more likely that those who do undergo resuscitation will experience an NDE on one of these occasions. The demographic changes underlying our experience of death ensure that hundreds of thousands of people each year will undergo these procedures. Consider the following observation by bioethicist Robert Veatch:[12]

Last year [1974] the world saw about fifty million people die. Because of the partial success of our new techniques of biomedical intervention, more and more of these people underwent a prolonged dying process. In the United States alone, perhaps a million people are now in the process of dying. . . . A casual look at American records of causes of death reveals a preponderance of lingering deaths, that is, deaths anticipated over months or years.

Life expectancy in all major industrialized countries, excluding indigenous populations such as Indians or aborigines, extends into the early seventies. Even among the indigenous populations in these countries, life expectancy is in the fifties. Although death from infectious disease is still rife in these latter groups, death from accidents and cardiovascular disease is also high. Cardiovascular events in industrial societies (primarily heart attacks and strokes), then, are still the leading cause of death today.[13]

The shift from farming and laboring to white-collar work characteristic of modern economies has produced a mortality profile that complements the lifestyle of these workers. Workers today are collected in large cities, spending the majority of every week working intensely, often under deadline pressures. Work-related migration is common, as is drug and alcohol consumption and sedentary leisure activities.[14]

By the very nature of these modern circumstances, the human circulatory system develops additional pressures from tobacco, alcohol, high-fat diets, and stress-induced adrenalin and cholesterol buildup at a time when it is least in a position to cope—the latter half of the life span. Thus the modern workaday world has been, with some variations due to cycles of economic boom and bust, largely associated with rising levels of cancer, heart disease, cirrhosis of the liver, and motor vehicle accidents.[15] Death from childbirth and infectious diseases has now given way to death from diseases of old age and the terrors of modern work such as the traffic jam, the boss, and the deadline.

In that context, resuscitation is not simply a possibility but a likelihood for most people. So, the large number of people who discuss their subjective experiences after revival are only one reason for popular coverage of NDEs. The experiences allegedly associated with them intrigue because of one's identification with those experiences, either in fact or as a possibility.

Nevertheless, the impact of resuscitation on many people who survive resuscitation must be balanced by the fact that few people are actually revived. Of those who are revived, fewer still live very long. Hermreck[16] estimates that the general survival rate is around 15 percent. He adds:

Virtually no patient with metastatic cancer, acute stroke, sepsis, gastrointestinal hemorrhage, or pneumonia who suffers cardiac arrest survives until discharge after CPR.[17]

Estimates do vary, but the figures are little better than those of Hermreck. Vrtis, for example, a nurse who reviewed several studies to ascertain the costs and benefits of CPR, found that in the hospital the resuscitation rate was 43 percent, with a survival to discharge rate of about 12 percent. The success rate for out-of-hospital CPR attempts is 16 to 19 percent of those who live long enough to enter the hospital, with only 3 to 8 percent surviving hospital discharge.[18] Follow-up studies show a further picture of deterioration. Vrtis reports three studies showing that of those patients discharged after CPR, 15 percent died within three months, 20 percent within six months, and 52 percent within three years. Significant neurological damage to the brain is another complication for people who survive resuscitation.[19]

However, technical reservations notwithstanding, most people are unaware of the poor prognosis of CPR and are drawn to stories of survival from genuine personal interest, as I have said, because they believe they may be in a similar position one day. In demographic terms, this is probably true. In terms of the general statistical probability of surviving CPR, their chances appear poor.

However, two reasons for interest still remain. First, even the small percentage who do survive nevertheless represent many hundreds of thousands of people around the world. The stories generated by this number of people, over time, must stimulate a considerable amount of interest from many quarters of society but especially from the lay community who remain ignorant of the efficacy of CPR. Second, during the period between CPR and discharge many may still discuss their NDE experiences with others, raising interest and stimulating a search for explanations more satisfactory than that of simple hallucination.

changing Attitudes toward Death and Dying

Those in the academic field of death and dying often remark that we are a "death-denying society," a people who are unable to discuss or deal with the horrors and sadness that death inevitably brings. Fortunately, this is neither true nor particularly revealing.[20] Because more than a few commentators on the NDE also implicate this attitude, I will devote the first part of the following section to challenging its relevance. I will then describe what I believe to be present attitudes toward death, which have greater value in explaining the popularity of the NDE.

Here are a few commonly cited examples of death denial: memorial gardens, current embalming practices, the rise of cryogenics, and the reluctance to speak about death or dying. A brief discussion of these will show that denial is more a poetic ascription than a revealing social one.

MEMORIAL GARDENS

Many observers have noted that these gardens no longer resemble cemeteries; rather, they are luxurious, parklandlike areas. However, their dissimilarity to past images of death does not mean a denial of death but merely a denial of those past meanings. Cemeteries were once places of fear and foreboding. They served to remind everyone that, whatever earthly glory that they now enjoyed, the material was fleeting, and they all ended as worm fodder in a field of stone and starkness. Today's cemeteries are designed to encourage bereaved people to return and to spend time reminiscing and reflecting. They are meant to attract people to return again and again. The former macabre images of judgment and isolation are meant to be replaced by modern images of rest and tranquility. But this replacement imagery can be no denial of death because this is, after all, a cemetery, and in such places one visits, consciously and unambiguously, the dead.

CURRENT EMBALMING PRACTICES

The image of lifelike corpses reclining in their velvet lined coffins was portrayed memorably in Jessica Mitford's *The American Way of Death*[21] and the film of Evelyn Waugh's *The Loved One*. However, whatever one's tastes are in this matter, one person's denial is another person's fashion sense. The fact is that the cosmetic industry has provided cultural images of beauty and health. It is uncommon for makeup to be considered a denial of poor health or ugliness. It is sometimes considered a denial of aging but, again, only when worn by the aged. This selective application of the idea of denial says more about the labeler than about the person (or body) selected for labeling.

It can be argued that face repairs for damaged corpses are probably more desirable than for some of the living. Death is a difficult thing for family and close friends to cope with; when violence is still present on the face of the corpse, this can be additionally traumatizing. Knowing this, embalmers attempt to minimize not the presence of death, which is literally undeniable, but the trauma associated with the image of violence.

RISE OF CRYOGENICS

Some people have paid extraordinary amounts of money to have their deceased friends or family members frozen. The idea is to preserve their bodies for a time in the hope that a cure for their cause of death might be found. They hope then to reanimate the deceased, provide the cure, and restore the person to his or her old life and company, thereby "cheating death." There are several points to make about this idea. First, given the media portrayals of medicine's successful war on disease, the idea in principle is not so strange. The confidence might be a bit overblown, but the rationale is reasonable in principle.

Second, cryogenics is not necessarily death-denying. When the purpose of cryogenic freezing is to reanimate a person so that he or she can live a life that was ended in youth or middle age, the idea is not directed at death per se but rather at premature death.

And finally, if the rationale behind some cryogenic practices is indeed to eventually achieve immortality through an endless process of freezing, reanimation, and cures, then perhaps some of the advocates of these practices are death-denying people. That possibility, however, no more characterizes our society as death-denying than the mere presence of Communists or hoola-hoopers makes us Communist or hoola-hooping societies. Societies are pluralist organizations. They contain a great diversity of membership, but a society is characterized poorly by the attitudes and practices of small minority groups such as those in the cryogenics movement.

RELUCTANCE TO SPEAK ABOUT DEATH OR DYING

For a long time, lay and academic observers have commented on the Western reluctance to discuss death. However, this reluctance is only partly true, and then only for some, at some times, and not for the reasons supposed.

In the thirty years preceding the 1960s, lay people often did display an extraordinary reluctance to speak about death, but these same groups found it just as difficult to discuss sex. Discussions of masturbation, homosexuality, incest, or child sexual abuse were thought to be inappropriate topics for dinner conversation. Indeed, the discussion of politics or religion, as quintessentially divisive issues, were also undesirable. These were regarded as controversial subjects.

But the people who felt this way about these topics were no more death-denying than sex-denying. Sex and death presumably continued as regular experiences for everyone during the 1940s, 1950s, and 1960s, but talk about them was privatized and organized by the sanctions of middle-class politeness. Death was not denied, nor was sex. Talking about them was simply a matter of bad manners. Sociologists Fulton and Owen discuss this pre–World War II generation of *Ed Sullivan Show* watchers as people whose origins were conservative and rural.[22] Concepts of death were derived from medicine or religion, and faith in these sources was strong. Thus, attitudes were certain and derivative. Communication about delicate matters was not always verbal. As Armstrong argues, in relation to death silence is not necessarily the opposite of the truth.[23] A look, a squeeze of the hand, and the collaboration of the dying in a game of mutual pretense and hope may have been a practice culturally appropriate to both parties at that time.

Their children, however, the postwar baby-boom generation, saw all these topics and experiences differently. By 1960, 80 percent of American homes had a television. Enrollment at colleges and universities has quadrupled since World War II. And it is estimated that the first television generation has viewed some 10,000 acts of homicide, rape, or related form of violence.[24] Even now, the sociologist Michael Kearl reports that, by the age of sixteen, the typical American has witnessed "some eighteen thousand homicides on television."[25] For these postwar generations, timidity about sex or death was not only laughable, it was also objectionable. The heart of American liberalism quickened in the 1960s, fueled by higher education, the Vietnam War, the rise of popular culture in music and protest, and religious reforms such as Vatican II.

And as the customs of one generation give way to those of another, it is also worth noting that professional discussion of death and dying had been increasing and gathering pace since the nineteenth century. Medical discussion began with interest in death certification, but soon extended to health and safety issues surrounding the corpse and its disposal. Studies of the problem of apparent death became "abundant."[26] And although the historian Aries[27] discusses the taboo against talking about death with the dying, talks with the family and other professionals continued unabated. During the 1960s and 1970s, this concern about death spread to the social sciences. Simpson[28] remarked that in the late 1950s there was almost no significant literature on the topic, but by 1976 there were over 750 books, hundreds of articles, and over 200 films.

We must conclude that the reluctance to speak about death is localized in time and population and that, where present, it is not confined to death but includes other areas of potential embarrassment. As an idea, denial is unhelpful in unraveling the mysteries behind that reluctance. This is because, apart from anything else, reluctance and willingness misrepresent the communication styles of the period and exaggerate the importance of death within them.

What, then, is the current attitude toward death and dying, and how can this attitude account, at least in part, for the popular response to the NDE?

CURRENT ATTITUDES

If American attitudes toward death before World War II can be called reserved, privatized, certain, and derivative, it is no coincidence that present attitudes are a reaction to them. The present attitudes toward death are characterized by four rather different social features, features forged from the material and social conditions of the baby-boomer generation. These attitudes are expressivity, uncertainty, individuality, and eclecticism. Each of these, as a cultural response developed during the 1960s and 1970s, make the NDE a celebrated and well-received part of popular culture today.

EXPRESSIVITY

Although the earlier generation was reluctant to broach the subject of death and dying, their children are making up for lost time. Indeed, expressivity, as the English sociologist Tony Walters remarked, is the hallmark of successful current grief work.[29] One must ventilate, express all the pain and anger for successful resolution of our fears, anxieties, and hurts about death and dying. Expressivity symbolizes the relaxed and the healthy, the high fiber of creative and constructive moods, attitudes, and feelings. It replaces the old view that expressivity is indulgent, attention seeking, rude, and precocious. Conversely, reserve is now considered awkward, withdrawn, difficult, and self-conscious. Reserve no longer indicates strength, acceptance, independence, or grace. From this culture of expressivity has grown a desire to hear from those who themselves confront death in any of its forms. The NDE, marketed as stories from those who have died and returned to tell the tale, is encouraged to do just that, again and again. This is encouraged by the current norm of expressivity because, apart from its value to others, it is thought to be beneficial to the speaker.

UNCERTAINTY

The sociologist David Armstrong reminds us that in the nineteenth century the body was thought to hold the secret of death.[30] Prior to that time it was the words of the dying person, which is why it was important for the dying and everyone else to know when dying commenced. When the "secret" of death was determined to reside in the body, an obsession with the causes of death began, first with medical certification and then with the practice of autopsy.

> . . . biological space was the repository of truth; beyond there was nothing and therefore no need to expend talk on spaces such as that between the dying and their entourage.[31]

But by the 1960s this certainty had given way to doubt, in fact uncertainty, first within the medical profession and later outside of it. The time of death, the cause of death, and the definition of death all became rather unclear as autopsy and pathophysiology practices challenged clinical judgment. Clinical death merged with biological death and blurred with cell death. Judgments became so difficult that the debate has moved from anatomy to physiology to ethics, where it currently resides. Out of that confusion, either as a way out of the maze or to bypass it, there has been a recent tendency to return to the patient and ask him or her about the subject. Those who have "died," such as the clinical NDEr, may be able to tell us about death since the academic debate appears hopelessly lost in the minutiae of conceptual hair splitting.

INDIVIDUALITY

We are no longer a culture where the individual measures his or her progress through life by the social stages or rites of passage in society. Many people now constitute single-person households, unable or unwilling to live with others. Many other couples live together but are unable or unwilling to have children. People marry more than once, have several jobs, have more than one career. The institutional norms for family, religion, and work have undergone a social revolution since World War II, and the philosophical liberalism accompanying these changes has encouraged the idea of personal choice.

"It's up to the individual" has become a moral and ethical imperative on issues ranging from women's health to film censorship. In that context, progressively more people are demanding access to information on experiences that might enable them to make such choices. Concepts of death are no exception. Medieval religious concepts of death appear vague or anachronistic. Medical conceptions appear uncertain or uncompromisingly materialist. Even the mystical conceptions of death associated with popular imported religions such as Taoism or Buddhism appear either oblique or unimaginable.

The image of death that emerges from the NDE is one of a detailed, organized set of experiences that, at least in content, are intriguingly familiar. The social images and ethics are generous and simple, and the unusual experiences can be understood by employing familiar parapsychological theories or assumptions widely understood by the general public.

Furthermore, the recent introduction of the subject is seen as originating from dissident medical authorities such as Kubler-Ross,[32] Moody,[33] and Sabom.[34] It therefore has an added, albeit unexpected, quality of authority, in contrast to being seen to derive from unconvincing religious or fringe cult sources. NDEs, then, have a genuinely "designer" quality for the discerning individual who is dissatisfied with the old ideas of death but is critical of reductionist material ones that, for example, emerge from the conservative ranks of medicine.

ECLECTICISM

As a direct product of the role of personal choice—derived, as I have said, from the modern tendency to practice existential browsing—much that appears attractive is eclectic. This is very much a postmodern development in the sense that, freed from a lineal view of progress and style, people now mix and match their material choices and their existential ones. House and building design in general, for example, now often incorporates styles or concepts from previous periods—Art Deco, Victoria, colonial, 1950s retro, and so on. Furniture and clothing reflect the resurgence of the 1960s revival or interwar fashion. None of these practices are concerned with reproduction and authenticity, but rather with a blending of styles suitable for the often casual work and leisure styles of the 1990s.

In matters more existential, such as birth, death, and marriage, people now commonly blend lay practices with religious ones rather than simply opting for one or the other. Music at these functions is just as likely to be by Willy Nelson or John Denver as church hymns from the eighteenth century. Functions are just as likely to be conducted by celebrants in chapels as by ministers in a forest grove.

The NDE is particularly suitable, and therefore attractive, for these purposes because it also blends well with all beliefs. NDEs are just what you would expect of a dying brain. Christians are able to show how the NDE is consistent with their own particular readings of the Bible about the afterlife or Satan's intractable and devious behavior. Spiritualists are able to breath a little easier, not only because of their reading of the paranormal aspects of the NDE narrative but also for glimpses of the Summerland often discussed by them. The NDE is the quintessential postmodern idea of death—eclectic in imagery; philosophically accessible to a wide range of beliefs without being particularly harmful to any of them; and critical of broad, singular, and simplistic ideas, whether materialist or religious. It is not an innocuous idea of death but rather a highly adaptable, and hence highly attractive, set of images. For religion, though, much of the NDE's integration has depended on the weakened ability of the clergy to ward off competitors, both philosophically and organizationally. This brings us to the next reason why NDEs attract so much popularity.

The changing Role of Religion

The appearance of the NDE in the popular press and media occurred in a period when present religious attitudes were experiencing significant social change. During the 1960s and 1970s, a new, highly educated cohort of churchgoers were beginning to ask many complex religious, moral, and social questions. How does the church view social protest, the Vietnam War, or homosexuality? What are the religious replies to social critics of religion such as Freud, who argued that religion was an illusion, or Marx, who argued that it was merely the opium of the people? What are the theological or doctrinal answers to modern biology's and psychology's claim that human beings are simply organic cyborgs, that there is no ghost in the machine? What, then, of God or a personal future beyond the grave for any of us? If God cares for us all, what happened at Auschwitz or Hiroshima?

Serious answers to these questions came slowly, if they came at all. People were offered ambiguous tracts of Bible text, told to pray for an answer, told to have faith, or had their inquiries downplayed or dismissed. This resistance or tardiness of the clergy in meeting the challenges thrown up by postwar social changes did not, somewhat surprisingly, lead to the wholesale abandonment of religion.[35] Rather, religious practice fragmented; its diverse membership spread out and sought philosophical alternatives that might be integrated into their current religious views or help them break with them. Because such intellectualism is often linked to organizational adherence, the rates of church membership, attendance, and belief in life after death began a steady but sure decline.[36] Some people abandoned religion altogether, but many more shifted their religious inclination to alternative religions such as Buddhism or spiritualism or even to secular philosophies such as Marxism.

Others sought what have been called *quasi-religions*—philosophies and movements such as astrology, transcendental meditation, and Scientology.

In this climate of religious change, the NDE held great appeal for those in alternative and quasi-religious circles because it seemed to deemphasize the importance of traditional indicators of spirituality: church attendance, doctrinal knowledge, rituals, public asceticism and tithing, and so on. The hallmarks of spirituality appeared to be a return to simple acts of kindness, self-reflection, and tolerance. Status-seeking materialism was eschewed in images of the NDE. The idea of the role or purpose of each life was reintroduced forcefully by NDE imagery. The universal relevance of these values and attitudes for all religions was appealing to persons outside traditional religions, those most likely to care less about denominational territorial disputes. For many who draw on Eastern religions and philosophies, it is the way such religions guide life in the present that is most important to them, rather than the images of the afterlife.

For those still adhering to traditional religions, the images of the afterlife suggested by the NDE may have been one important influence in the popularity of NDEs for this group. This is because traditional images of the afterlife seemed decidedly tired and unconvincing. Many of these images were either vague, depressing, or surreal. Rabbi Malino[37] provided a timely review of these rather hopeless images in the mid-1960s. For adherents of Judaism, people who died could look forward to a life of sorts in Sheol, a rather charmless, gray place where they existed in an unconscious or stupefied state. Malino observes:

> God's spirit could bring no animation to the countless generations doomed to an everlasting quiet . . . it was an empty, unenviable eternity devoid of meaning and irrelevant to life.[38]

The Bible has very little to say about the afterlife, as it is preoccupied with encouragements, warnings, supplications, observances, laws, parables, and prophecies about life. This was a rather disappointing omission, from the Christian point of view, and gradually saints supplied what the Bible chose to be modest about.

In the medieval period, detail on the afterlife developed to an extent that would impress even a seasoned ethnographer. Heaven was the place where the righteous would reside after death, a place where God, his angels, and the saints also happened to live. Purgatory was a reasonable place where borderline but fairly good people went, to be purged of their sins, to be purified and made ready for their eventual ascent to heaven. And hell, of course, was a sad and bad place where Satan ruled. Punishment was eternal, a combination of endless torture and suffering induced by the fact of separation from the glory of God and the sadistic practices of the devil and his imaginative cronies.

Climatically speaking, hell was dark and very hot, while heaven was full of golden light, blue sky, and cumulus nimbus clouds. Such images have achieved longevity through the iconography of European Catholicism. These images were exported on a million holy cards, scapulars, medals, and church artwork. Together they renewed each generation's acquaintance with their scenes up to the 1960s, when changes flowing from the Council of Vatican II began slowly to do away with them.

Protestant images were rather vague but, for that reason, somewhat more cheerful than those of Judaism or Catholicism. Protestant artwork emphasized pastoral images where heaven was an endless walk with Jesus. Hell, however, particularly for fundamentalist churches, encouraged belief in images just as powerful and frightening as any in the Catholic imagination. In that respect, all Christians were made to feel unsure and rather fearful about their afterlife prospects.

At a time when theological concepts such as the afterlife, the soul, and divine grace and intervention appeared outdated, even ridiculous, attempts were made to develop other ways of seeing immortality. Malino, for instance, suggests that one might live on through children, the thought and memory of others, the influence we have on others, or through identification with "the timeless aspirations of the spirit."[39] Such was the poverty of afterlife imagery for all major churches in the West that now they retreated from its scenes altogether. Suddenly, into this low ebb of the religious imagination, stories about the NDE appeared, and from a most unexpected quarter medicine.

The NDE was characterized by six features that made it attractive to members of traditional religions and to those who believed in life after death but didn't know what that actually meant for them. First, most of the NDEs reported by Kubler-Ross and Moody in the first few years of the NDE's public introduction were positive experiences. Negative NDEs were either denied or doubted. Those that were reported were associated with overtly Christian interpretations. It was not until the appearance of Margot Grey's work in the mid-1980s that negative NDEs were recognized and taken seriously.[40] This gave the image of the afterlife portrayed by the NDE, at least initially, a very positive and attractive appearance.

Second, the life review in the NDE was *not* a medieval version of divine judgment.[41] If this was the beginning of an afterlife, the life review appeared to function as a learning experience for the departed. The "lessons" so gained were to be of some use in further self-development on the other side. This was also an image of hope and added, I think, to its considerable appeal.

Third, the light accorded wonderfully with biblical references to Christ as the "light and the way" and to many other references in which light is central to religious images of hope and warmth. The descriptions of the light's behavior appeared comforting and attractive; it was forgiving, kind, gentle, even funny!

Fourth, images of "heaven," if this is how some wished to interpret some of the social scenes, were familiar. Heaven did not appear foreign, however pleasurable. The physical, geographic, and social images were an extension of life here. Furthermore, life there could eventually continue with the friends and family known here. Death would be reunion.

Fifth, all these images and ideas removed the old ideas of death as something dark and cold, with the pallor of sin and the uncertainty of judgment hanging over each traveler to the next world. The medieval image of a struggle between God's angels and the devil's henchmen for the luckless and vulnerable soul of our departing spirits was banished by the powerful image of a possessive and generous being of light.

Finally, and most important, the NDE put the idea of personal survival after death back on the religious agenda—not as something vague and ambiguous, not as a mixed blessing, but as something to understand and look forward to. The elite of many churches have argued that the details of the afterlife, if any could ever be discerned from any source, should be of little concern to a Christian. This is because faith should be the cornerstone of the religious imagination, not cheap promises of the good life as a reward for deeds performed here.

But images of the afterlife, whether from NDEs, spiritualists, or the Sistine Chapel, do require faith, so that admonition is very much beside the point. The idea of an afterlife is not substantially different from the idea of God or divine purpose; they are all important sources of hope and inspiration. And as Zaleski[42] has shown, such images change from one period of history to another as each generation's understanding of itself and its hopes changes. If there is a consistent meaning to the religious idea of revelation, this must surely be it. Each generation reinterprets its understanding of divine and human purpose based on its own spiritual and social experiences. Political and military history easily shows that this does not necessarily translate into desirable behavior. Promises of the afterlife, cheap or otherwise, cannot and have not, of themselves, encouraged deeds good or evil. This is because eschatology is not a substitute for a philosophy of ethics, nor is it meant to be in any religious tradition. Nevertheless, for some people, the

NDE has supplied both a simple ethics and an important source of religious hope. It has restored the individual and hope to their place in religious eschatology, to be discussed and debated in both alternative, quasi, and traditional religions. For this unexpected source of renewal, the NDE has been attractive and popular to most religious people everywhere in the Western world.

Changing Relationship with Social Institutions

There has been growing cynicism about the beneficence of our social institutions: our ability to effect change within them and their ability to deliver the guidance and services that they promise. We have seen how religion, to some extent, has been affected by this attitude change. However, religion is not the only social institution affected. There has been widespread disenchantment with established organizations in general, and this, in turn, has provided fertile ground for the growth of a variety of contemporary social movements. Christopher Lasch, in the preface to his best-selling book *The Culture of Narcissism*, states the problem this way:

> The inadequacy of solutions dictated from above now forces people to invent solutions from below. Disenchantment with government bureaucracies has begun to extend to corporate bureaucracies as well—the real centers of power in contemporary society. In small towns and crowded urban neighborhoods, even in suburbs, men and women have initiated modest experiments in cooperation, designed to defend their rights against the corporations and the state. The "flight from politics," as it appears to the managerial and political elite, may signify the citizen's growing unwillingness to take part in the political system as a consumer of prefabricated spectacles. It may signify, in other words, not a retreat from politics at all but the beginnings of a general political revolt.[43]

These remarks are echoed in John Naisbitt's analysis of contemporary American social trends entitled *Megatrends:*

> For decades, institutions such as the government, the medical establishment, the corporation, and the school system were America's buffers against life's hard realities—the needs for food, housing, health care, education—as well as its mysteries—birth, illness, death. Slowly we began to wean ourselves off our collective institutional dependence, learning to trust and rely only on ourselves.

During the 1970's, Americans began to disengage from the institutions that had disillusioned them and to relearn the ability to take action on their own.[44]

The social attitude now on the ascendant is the "self-serve, self help" view of the world. Home births, deaths, and marriages; the self-sufficiency movement; alternative and preventive medicine, food, and welfare coops; community crime watches; consumer protection societies (including medical consumer groups); and home education groups are just some of the examples of the trend toward self reliance.[45]

Even where established social institutions are patronized, the social and sometimes political organization of these forums have changed. Lay participation in the practices of these institutions has widened, with people now actively engaging in the central aspects of religious rituals (as lay ministers and priests) and in life events from birth (father's participation) to death (in hospice care). This participation has also been political. The insistence on outside representation on management committees, particularly from consumer groups, has strengthened. Overall, our former trust in institutional help and advice has become conditional.

The reasons behind this altered relationship with our key social institutions are many. Some have argued that the rapid social changes brought about by the uneven and inequitable spread of industrial capitalism in this century have been destabilizing, making modernity a fertile breeding ground for radical criticism, cultural experiment, and alternative lifestyles.[46]

Others have argued, more simply, that the increasing industrialization and technological growth in all our lives has led to a counter social reaction. The more impersonal the world grew, the more personalised additions or alternatives were forged to meet, complement, or reject these high-tech developments. People became more interested in the personal, social, and spiritual just at the point in their organizational development when the technical, technological, and functional imperatives seemed at their height. There was a desire to reunite the subjective and objective aspects of our lives just at the point when they were perceived to be separating. As Naisbitt observes, in the movie *Star Wars:*

When Luke Skywalker flies in on that final run, the Force with him, he turns off his computer, but not his engine.[47]

Still others have sought to place responsibility on an increasing preoccupation with the self as the latest, perhaps final chapter in the rise of bourgeois individualism. Unable to see a way through the complexities of corporate bureaucracy and nuclear age politics, people now look inward toward the self, in therapy, public confession, Eastern religions, or psychic communications. In this twist of feeling and priority, the person of today sees his or her internal world as the only world of realization and possibility, feeling only impotence and distrust toward public institutions and their pronouncements.[48]

Whatever the reason, no major institution has escaped these dissatisfied and restless publics. For the NDE, this has meant an inability of institutional authority to dismiss it easily. This is because the NDE, especially the clinical NDE, represents the reemergence of the personal amid the seemingly impersonal nature of modern death—from the sight of nasal tubes, oxygen masks, and white lights and uniforms to the beeping and jarring sounds of electrocardiogram machines and defibrillating procedures.

The message of many NDEs, although highly personal in experience, is commonly social rather than religious and critical rather than affirming of present social conditions and attitudes. The emphasis on self-development, service to others, and learning for life rather than for functional or career reasons fits snugly with the current preoccupations with self and purpose.

And finally, NDEs signal a new way to learn about the nature of death: we can learn from the personal experiences of others, however clergy or doctors explain it. The personal can triumph over the academic because technical explanations are always partisan and factionalized. One's own explanation, in the final analysis, is at least true to one's own intellect, feeling, and experience, and that bias is at least accountable to oneself. One knows the judge and the accused well.

On the other hand, the new cynicism toward authority, which stems from public awareness of competing views and factions within all institutions, means that people approach such judgments very cautiously. The anxiety generated by controversial subjects, from uranium mining to the safety of silicon breast implants to NDEs, is not quieted by the mere presence of "scientific" reassurance. People want to know *whose* scientists are doing the reassuring.

The belief in the value neutrality of science is long gone, together with unconditional reliance on the local doctor, lawyer, teacher, and parson. From medicine, for example, the NDE has attracted more than a few explanations, but the authority of these has been feeble. As I indicated earlier, surveys still show people preferring life-after-death explanations over medical ones. Clearly, medical authority has taken a battering in the last twenty-five years—not simply about matters regarding death, as I will discuss later, but also more generally.

Social sciences have made trenchant criticisms about medicine,[49] but doctors have rarely taken much notice. Others have adopted a different attitude.

Journalists and other social commentators have been summarizing and translating some of this criticism to provide explanations for a very dissatisfied and growing public in search of answers. This has contributed to widespread cynicism about medical authority, reducing much of its credibility on a variety of fronts and topics, including the NDE. The medical response to the public's rather cool reception has not helped medicine either. We will examine this issue in detail in Chapter 7.

Overall, it must be concluded that the changes in our relationship to institutional authority since the 1960s have encouraged most people to give unusual personal and social experiences a fair hearing. People are not ignoring the established authorities, such as medicine, but neither are they easily persuaded, as in the past. The test of an authority's credibility now rests on its local success: the program worked or didn't for my neighborhood; the cure worked or didn't for my partner; the technique saved my daughter's life or it didn't; the explanation makes sense to me or it doesn't. More Americans than ever before have a higher education, and that cultural fact has changed the game—not simply for television marketing but more fundamentally in the politics of credibility and the social consequences of professional insularity and nonaccountability.

conclusion

The simple belief that the popularity of the NDE has to do with the appeal of immortality suggested by some of its imagery is an oversimplification, to say the least. Furthermore, it ignores the fact that the immortality promised by traditional religions has not helped stem their falling membership or reduced attendance. Nor has that promise prevented other people from developing an interest in Eastern religions whose images of the afterlife do not necessarily cater to the idea of individual survival after death. Finally, the idea that the crowd-pleasing feature of the NDE may be the promise of immortality reduces the complexity of the NDE to its crudest level. People are attracted to the NDE for a number of its characteristics, and for a number of social and historical reasons.

In this chapter, we have explored some of the principal reasons for the popularity of the NDE: the changing experience of death due to rising life expectancy; advances in resuscitation techniques; the impersonal nature of modern death; the growing interest in matters concerning death; changing attitudes toward former social taboos; dissatisfaction with traditional religions; growing skepticism toward institutional authority; and the consequently rising interest in personal testimony and experience.

All of these reasons play an important role, each of them laying the social groundwork that makes the NDE appear relevant to the anxieties and questions of the day. And that relevance is the key to understanding the continued interest in what appears to be merely a footnote in the annals of contemporary academic medicine.

But the reasons discussed above are not the only social influences behind the unique attracting power of the NDE. I believe there is one more important reason why NDEs continue to intrigue us. This reason is found in the inspirational images of another society, experienced briefly by some NDErs. These are given wide publicity in the press and significant space in more than a few of the best-selling NDE accounts. These are not necessarily viewed by many as scenes of heaven, but rather as images of an ideal society.

6

IN PURSUIT
OF THE IDEAL SOCIETY

For most of this century, the social science literature dealing with death and dying has been explicit about one issue: Death is a dark country. Never far away, its major cities are Loss, Grief, and Aloneness. Recently, however, a new viewpoint has emerged. Beyond that first shadowy country lies another, less inhospitable land. Indeed, this is a land of fabulous light and landscape. And in this country the cities are called Learning, Love, and Service. This is the place many people describe when recovering from the NDE.

Less astonishing but equally intriguing is the curious paucity of sociological literature on this strange society. Is this society, and its cities and citizenry, the latest modern notion of paradise? Do these visions and values of the good life bespeak a renewed desire for some lost Arcadia or golden age? Or do these visions in the final moments of consciousness reveal, at death's door, a final yearning for utopia?

The aim of this chapter is to examine the transcendental features of the NDE that depict a social world beyond this one. As usual, I will not attempt to evaluate the reality of these reports. My purpose is to treat the reports of and writings about transcendental NDEs as narratives that may be read and interpreted for their assumptions about and allusions to the ideal society.

In this way, I argue that NDEs, whatever else they may be, are social images that belong to the historical and social discourse about the ideal society. The identification and examination of this otherworldly society permit a cultural analysis that furthers our understanding of NDEs as sociological phenomena. If visions of this otherworldly society are prompting people to change their social values and lifestyles, then it is important to understand why. As we have seen in our examination of shipwrecked castaways, many of these changes undoubtedly derive from the social and psychological crisis of being near death. However, another part of this understanding must come from the possibility that the social images of this ideal society may also be prompting or inspiring some of these changes. Furthermore, treating the NDE as part of a discourse about the ideal society makes it possible for us to understand one further possible reason for the popularity of NDEs. Perhaps the social imagery in NDEs can help us reevaluate our social ideas about personal identity, social change, deviance and control, and issues of cultural and social representation. In exploring the transcendental features of the NDE, I attempt to answer two questions. First, what is the nature of this ideal society that so many NDErs encounter? Second, how does this conception of the ideal society differ from earlier ones?

In developing answers to these two questions, I organize the chapter in the following manner. The first section will identify the type of society that people report seeing in the NDE. I call this society the *transcendent society*. I will then describe the social features of this society as revealed in various NDE accounts. The next section will compare the features of this society with five types of ideal society as outlined and discussed by J.C. Davis.[1] The final section will discuss the transcendent society as a utopia with unique social properties. I argue that, as a utopian form, the transcendent society reawakens the pursuit of the ideal society. This is a pursuit that has largely faltered in this century because of several historical and conceptual problems. Many of these difficulties are overcome by the transcendent society, and perhaps this is a key influence accounting for the attracting power of the NDE, particularly for the general public.

The Transcendent Society

The overwhelming majority of reported NDEs are positive experiences. Bruce Greyson identified three distinct types of positive NDE.[2] These are the cognitive type, which exhibits time distortion, thought acceleration, life review, and sudden understanding; the affective type, which exhibits feelings of joy, cosmic unity, peace, and an experience of light; and the transcendental type, which exhibits encounters with an unearthly realm populated by beings and a "barrier or point of no return" that, if crossed, would preclude a return to life. It is the transcendental type of NDE on which this chapter focuses for details of a society beyond death.

Greyson estimated that over 40 percent of his sample of NDErs experienced the transcendental type of NDE. Michael Sabom, who divided the NDE into two types, autoscopic and transcendental, estimated that over half of his sample of NDErs had encountered another social world beyond this one.[3] More modest figures were reported by the Evergreen study (34.5 percent)[4] and by Kenneth Ring (20 percent).[5] Furthermore, in 1982, George Gallup estimated that some 8 million Americans may have experienced an NDE.[6] Many of these persons are presumably familiar with aspects of the transcendent society.

In the NDE literature, descriptions of the transcendent society are often the spectacular finale in an ideal composite portrayal of the NDE. Researchers such as Moody,[7] Ring,[8] Sabom,[9] Grey,[10] and Zaleski[11] developed their analyses by moving from the basic cognitive-effective features of the NDE to the descriptions of an unearthly realm. As with many features of the NDE, the reports of this society are highly similar. However, the details of the transcendent society are admittedly few. As Zaleski noted, the emphasis in most descriptions is on the message of love, learning, and personal and social transformation. However, sparse as these details may seem, it is still possible to discern salient features of organization and process that would locate this kind of society in the context of others.

Lundahl provided the most systematic social and physical description of this other world, based on his review of the NDE literature and on nine selected accounts of Mormon NDEs, but these details are not confined to his report.[12] The physical world in the NDE is a world of beautiful skies and lush vegetation. Crisscrossed by streams, dotted by lakes, it also features forests, lawns, parks, and gardens that contain flowers of unique and unprecedented beauty.

> I can remember it all even though it was a long time ago. I was in a big garden and I could see a stream and trees laden with pomegranates. It was extremely bright and there were people there.[13]
>
> I don't know how I got there, but I found myself in a beautiful country lane. I was strolling down the lane slowly and I felt I had all the time in the world. I

could hear the skylarks singing and I thought "Oh, how lovely." I particularly noticed the colors; the sky was a brilliant blue, but the colors were so soft. The green of the trees, too, was brilliant but not harsh.[14]

As I went outdoors into the garden I saw mountains, spectacular valleys, and rivers in the distance.[15]

Sight, movement, and mental abilities are increased, allowing greater vision, faster travel, and sharper cognitive skill. Although there are rural environments, there are also cities and many buildings such as halls, houses, and temples. Cities contain libraries, places of higher learning, and living areas.

I was taken to another large room similar to a library. As I looked around it seemed to be a repository of knowledge, but I couldn't see any books. Then I noticed ideas coming into my mind, knowledge filling me on subjects that I had not thought about for some time—or in some cases not at all. Then I realized this was a library of the mind.[16]

I was given a choice then as to whether I should go across to the city or whether I should go and have a look at some archives.[17]

We entered a studio where music of a complexity I couldn't begin to follow was being composed and performed. There were complicated rhythms, tones not on any scale I knew. "Why," I found myself thinking, "Bach is only the beginning!"

Next we walked through a library the size of the whole University of Richmond. I gazed into rooms lined [from] floor to ceiling with documents on parchment, clay, leather, metal, paper. "Here," the thought occurred to me, "are assembled the important books of the universe."[18]

And then I saw, infinitely far off, far too distant to be visible with any kind of sight I knew of . . . a city. A glowing, seemingly endless city, bright enough to be seen over all the unimaginable distance between.[19]

The social climate is described by Lundahl as largely one of contentment, happiness, harmony, and order. It is highly organized and eschews disorder and confusion. Apparently, people in this society work; that is, they have occupations that are often tied to some sort of human service industry.

Whatever else these people might be, they appeared utterly and supremely forgetful—absorbed in some vast purpose beyond themselves. Through open doors I glimpsed enormous rooms filled with complex equipment. In several of the rooms hooded figures bent over intricate charts and diagrams, or sat at the controls of elaborate consoles flickering with lights.[20]

As we approached the people, I saw that they were weaving on large, ancient-looking looms. My first impression was "how archaic" to have manual looms in the spirit world. Standing by the looms were many spiritual beings, male and female, and they greeted me with smiles. They were delighted to see me and moved back from one of the looms to let me have a better look. They were anxious for me to see the workmanship of their hands. I went closer and picked up a piece of the cloth that they were weaving. . . . The workers explained that the material would be made into clothing for those coming into the spirit world from earth.[21]

Interaction is based on cooperation in general, but sanctions do exist to control deviance.[22]

The social system is stratified, apparently along moral lines. The society is divided into different communities or levels of activity and order based on different degrees of moral progression. Problem groups are confined to certain areas so that they are restrained from disrupting the smoother operation of other communities.[23] The means and criteria by which such people are restrained are not clear.

Extraordinary as this description of the transcendent society must seem, it is incomplete for a reason other than the brevity of most NDErs' visits. Other societies with similar characteristics, or other cultural communities within this one community, also apparently exist. As we have seen earlier, there are suggestions of a transcendent society in Melanesian, Indian, and Chinese NDE accounts. Recall, for example, that in the Melanesian version of NDE, the society beyond was also beautiful and well ordered. No cities were reported; instead, there were descriptions of villages. In these villages people were also described as working, constructing buildings, or participating in traditional song and dance. Social control is also part of the work of this society; Dorothy Counts reported the trial of a sorcerer.[24]

In most cases, NDErs report their reluctance to leave that well-ordered world beyond. As some of Ring's respondents expressed it:

"The most depressed, the most severe anxiety I've ever had was at the moment I realized I must return to this earth."

"I began to realize that I was going to have to leave and I didn't want to leave" [begins to cry].[25]

Not all NDErs feel this strongly about their return. However, even those who desire to return often depart with a sense of regret because they are leaving such a beautiful place. The extraordinary nature of their story of revival and their accounts of another realm have fired the imaginations of millions of non-NDErs.

No doubt this description of a society is an ideal, indeed idealized. The sociological question is: what kind of ideal society is this?

Davis's Typology of the Ideal Society

In theoretical terms, the literature examining utopias and other forms of the ideal society falls into two categories. On the one hand, some writers are reluctant to attempt to define such societies for fear of excluding some types. Frank and Fritzie Manuel[26] avoided definition of major concepts such as *utopia* for fear of obscuring what they argue to be the pluralist nature of utopia.[27] This argument is similar to that of Krishnan Kumar.[28]

On the other hand, writers such as Davis argued that definition is not only possible but desirable precisely because of the ambiguity of the concept.[29] To this end, Davis provided a typology of ideal societies common in Western history. Davis maintained that definitions are important for clarity but that they do not have to be distinct and mathematical in construction; they do not have to be airtight. My view of this debate is that precision rather than definition is problematical. Karl Mannheim provided a precise distinction between utopian and ideological forms of consciousness.[30] Utopias are ultimately realizable ideas and programs; ideological ones are not.[31] The problem with this view is that we must wait until the end of history to identify which is which, a situation that questions the usefulness of the criterion.[32] Davis's typology allows us to identify and understand types of ideal society by their commonly occurring features. They are, in Max Weber's sense, "ideal-typical" categories; that is, they are approximations that permit variety.[33] They are not intended to pinpoint but rather to guide our thinking about the history of ideas concerning the ideal society.

Davis outlined the social and political features of five types of ideal society: cockaygne, arcadia, moral commonwealth, millennium, and utopia. I will examine each of these in turn and assess their applicability to the society reported in transcendental NDE states. Our review will demonstrate that past ideas about the ideal society are present in contemporary NDEs but that only utopian images seem to best reflect the types of societies portrayed in those visions.

COCKAYGNE

The cockaygne society has been described as the "poor man's paradise." This society exists in idyllic physical surroundings with material privileges and sensual gratification. Every whim and appetite is instantly and handsomely satisfied. A desire for food is immediately met by banquets of desirable items, which may be chosen from overloaded and groaning tables or placed in one's mouth without effort. Sexual desire is catered for by the instant appearance of beautiful, willing, and most able partners. The "Land of Cockaygne" was a prominent set of ideas about the ideal society in late medieval Europe.[34]

Davis cited a marvellous poem from the period that amply illustrated the spirit of cockaygne:

> *Ah, those chambers and those walls!*
> *All of pasties stand the walls,*
> *of fish and flesh and all rich meat,*
> *The tastiest that men can eat.*
> *Wheaten cakes the shingles all,*
> *Of church, of cloister, bower and*
>   *hall.*
> *The pinnacles are fat puddings,*
> *Good food for princes or for kings.*
> *Every man takes what he will,*
> *As of right, to eat his fill.*
> *All is common to young and old,*
> *To stout and strong, to meek and*
>   *old.*

The poem ends with the final message about work and payment:

> *Every man may drink his fill*
> *and needn't sweat to pay the bill*[35]

In cockaygne, the vision of ideal living is largely an escapist peasant one. In this respect, freedom from work and hunger are the main obsessions. The medieval social order is reversed in cockaygne, where peasants enjoy unrestrained decadence and the upper classes toil chin-deep for years in dirt and filth before they are able to indulge in any pleasures. The social order is maintained because appetites are always satisfied rather than because the basic organization of society has altered.

The transcendent society has some environmental similarities with cockaygne. It is a place that is perpetually beautiful and gardenlike. In the transcendent society, as in cockaygne, there is no death. However, in cockaygne, total wish fulfilment prevails.

Although many needs are apparently gratified in the transcendent society, there are no reports of cockaygne-like indulgence. In fact, except for the occasional report of fruit on trees, food appears conspicuous by its absence.[36] Also difficult to locate in modern accounts of NDEs are any descriptions of sexual activity. If eating and sexual activity were as prominent in the transcendent society as in cockaygne, the difficulty of locating these accounts would be unusual. Finally, the cockaygne life is an idle one, without work or desire for it. The transcendent society is one, however, where work is a conspicuous feature of social life. Buildings and service to others do not just happen but are provided for by fellow beings who fully plan and participate in the processes of this work. Clearly, the transcendent society is no cockaygne.

ARCADIA

The arcadian society is a cockaygne-like society with restraint. Set in idyllic surroundings once again, human beings display their dignity by exercising their moral and aesthetic senses. This restricts appetite, and so abundance is temperate and somewhat modest in comparison to the cockaygne. In a society of plenty, people do not overindulge but rather satisfy their needs. We can also observe this feature in the transcendent society. Grey, for example, cited the meeting of one NDEr with his deceased mother.

> I found myself standing in front of a nice prefab (inexpensive and prefabricated dwelling that can be erected very quickly and was extensively used during World War two to house bombed out victims). There was a path leading up to the front door with masses of nasturtiums on either side. The door was open and I could see my mother inside. I thought, "That's funny, my mum always wanted a prefab and she always loved nasturtiums."[37]

It is not important here to ascertain in whose mind the prefab/nasturtium existence was truly ideal, the NDEr or his mother. It is sufficient to note that basic desires are being fulfilled in this transcendent society. People's needs are being met, but not disproportionately to their desires, and their desires seem appropriate to their former lives and backgrounds. However, once again, the arcadian existence, largely inspired by romantic medieval notions of primitive life in the New World, is an idyllic life.[38] Work is not an integral and socially important activity for arcadia, and yet it is for the transcendent society.

PERFECT MORAL COMMONWEALTH

In this type of ideal society, people apply more restraint to themselves, tolerate some hardship, and work for the greater good. The idea of the moral commonwealth embodies the first major social shift away from the self and its needs to the needs and welfare of the wider community. The notion of regulation is formally introduced as an integral and important part of this ideal society. As its name implies, the wealth and work of every individual must be dedicated to the common good.[39]

The philosophy and sociology of the moral commonwealth turns on the idea of moral individualism. This society does not depend on structural reorganization but rather on the willingness of individuals to do their duty. Moral rearmament of the individual rather than changing the political and social systems brings about the ideal, harmonious society. Control and regulation exist within the individual. The problems of evil and deviance are inextricably bound up with the problems of personal discipline and values.[40]

In the NDE, the values of personal change and moral development are encouraged in the transcendent society in a context of social support, human warmth, and love. Nevertheless, the transcendent society has apparently not left its organization dependent on the combined efforts or willingness of individuals to maintain its order. Social regulation is clearly evident and, unlike some examples of the moral commonwealth,[41] magistrates are not redundant.[42]

As described by Lundahl, formal sanctions are enforced, and although there exist many "self acting and self thinking" individuals, more than a few people are restrained, presumably against their will.[43] A governing order that exerts some authority and control acts as a regulatory social system. This is clearly a system that takes deviance for granted and does not expect uniform moral restraint by all individuals, at least initially. The moral commonwealth model of ideal society, dependent as it is on individual moral restraint, does not appear to be the basis of the transcendent society.

MILLENNIUM

The ideal society of the millenarians is one where both human beings and nature are transformed by external apocalyptic forces. After purging of the manifold problematic features and groups within humanity, a new world order emerges.[44] Linked closely to the history of Christianity, the apocalyptic event is commonly Christ's second coming. However, the millennium applies to any religious movement and is an ideology of salvation that stresses perfection on earth facilitated by supernatural beings.[45] In social terms, the transcendent society is least like the millenarian society.

First, the appearance of the transcendent society is not linked to any ideology of salvation. That is to say, entry into that ideal society is not dependent on membership in any religious movement. Second, perfection on earth is not stressed. The transcendent society is an order that exists beyond, but alongside, our own and does not represent a future transformation of our own time and place. Furthermore, the transcendent society does not itself assume human perfection, as my earlier remarks about deviance and control suggest. Rather, moral and social evolution is assumed to be a process that may begin on earth, definitely continues in the transcendent society, and may be completed there in some distant and unclear time and place. Finally, the role of supernatural beings in the transcendent society is different from that in millenarian conceptions. Some NDErs do observe, or believe they observe, religious figures in the transcendent world, but these are often simply guides, life review facilitators, or part of a welcoming party. Some apocalyptic changes have been forecast for our world and its societies,[46] and some of these have suggested a new emerging order. This order, however, has never been confused or identified with the transcendent society of the NDE. That society apparently remains distinct and otherworldly.

The transcendent society is not a millenarian society, though some who visit there may bring the occasional millenarian message. Millenarians search for meanings in personal experience and in the world around them, like everyone else. But unlike many others, millenarians often find this meaning in ideas about the afterlife, in what John Harrison described as "holy utopias."[47] The path to this utopian vision is through apocalyptic and supernatural intervention. For the NDEr, the path to the ideal society is simply and less dramatically through death. Nevertheless, the paths of millenarians and NDErs may often converge, even when their purposes and ideas are not always the same.[48]

UTOPIA

According to Davis, utopias may be distinguished from other forms of ideal society by their approach to the problems of human willfulness, deviance, and unlimited appetite. In utopia there is no wishing away of problems, as in cockaygne. Nor is there at the center of utopian vision and planning the need for a great purging, as in the millennium vision. Rather, social systems must be designed to take account of social problems such as crime, hostility, and exploitation. Organizations must provide education and social control to enable the collective to attain greater good and harmony, but also to check the incorrigible, the corrupt, and the slack. As Davis observed:

> The perfect moral commonwealth tradition idealises man [sic]. The land of cockaygne idealizes nature, in an admittedly gross way. In Arcadia, too, nature is idealized but at the same time man is naturalized. In utopia, it is neither man nor nature that is idealized but organization. The utopian seeks to "solve" the collective problem collectively, that is by the reorganization of society and its institutions, by education, by laws and by sanctions.[49]

Utopias are total physical and social constructions, total environments, whose goals are social order and the ultimate perfection of humanity through collective effort. This is a good description of the transcendent society revealed in many NDEs.

As Lundahl has remarked, the transcendent society emphasizes harmony and organization. Order in both moral and social terms is the hallmark of that society. Education and social control of deviance are common. Education, as a formal way of gaining knowledge, frequently takes place in classroom-like settings similar to those of our own societies. Learning, in both formal educational and informal socialization senses, is an important feature of the transcendent society. As in most utopias, people are free agents, but this is a *freedom from* disorder and moral chaos. The citizens of utopian societies are not *free* to do as they please if this means creating confusion or doing wrong to others.[50]

Another way in which these values and images of the transcendent society are utopian is their function as social criticism. Utopian thought is always, partly at least, a rejection of the contemporary world and its processes.[51] It contributes to a "climate of opinion" that stimulates others to take up the policy, theory, or social action.[52] In this respect, the basic organization of the transcendent society conveys value systems that are utopian. Values important to NDErs, such as cooperation, humanism, and self-development, are an implicit criticism of other values such as competition, selfishness, and authoritarianism. Utopian values display and highlight rather than supply a specific outline of a new morality. They are inspirational rather than prescriptive.

The transcendent society and its tales, act as narratives by which we may orient ourselves, our cultures, and our roles and ambitions within them. In these ways, utopias are to adults what fairy tales are to children.[53] They draw on current feelings and problems about the world and inspire both audiences to higher things without ever becoming a dense legislature. This is a commonly observed role of utopian imagery.[54] This inspirational role makes utopias responsible for introducing or renewing a new and better set of human values. In the case of the transcendent society where the values are simply learning, love, and service, the task may arguably be one of renewal and revision.

Harmony, cooperation, and love are the chief characteristics of social intercourse in the transcendent society, as indeed they may be with NDErs themselves. In these respects the transcendent society, or what little we know about it, seems to meet Davis's general criteria for a utopian society. In these terms, the transcendent society is a total physical and social environment whose goals are human perfection, social order, and harmony. Furthermore, like Thomas More's Utopia, it is a society supposedly already in existence rather than being a prescriptive or futuristic entity. In this respect, the transcendent society belongs to that tradition of literature where a person or group of travelers stumble by accident on another society. Their accounts simply describe what they see, experience, and do, along with their incredulity and admiration.[55] In this tradition of utopian literature, these places often have cities of "a structurally fabulous kind," "miraculous transport and strange animals and people."[56] The transcendent society is characterized by many of these features too, but it is not simply a utopian society, like many before it. On the contrary, the transcendent society is a utopian society with several unique features.

A Unique Utopia

If the transcendent society is utopian, it is no ordinary utopia within the strict terms by which Davis outlined his typology. It is true that the social organization and values of the transcendent society are utopian in their regulation of work, deviance, and education. In this respect, social organization is the chief agency of socialization and control. However, the need for individuals to take responsibility for shaping and nurturing their own values is also strongly present. This is, for example, a common idea running through NDErs' reviews of their lives both during and after their NDEs. This social dimension of the transcendent society is somewhat akin to the prescriptions of the moral commonwealth.

Although utopia's main task is the transformation of humankind, nature also seems perfected and idyllic in the transcendent society. Here we witness elements of cockaygne. But restraint does coexist with the satisfaction of a wide array of needs; in this respect, there exist elements of arcadia. Although the transcendent society is not millenarian, millenarian elements do overlap, particularly the pareschatological direction and dynamism that NDErs and millenarians draw on for their images.

So the transcendent society is utopian but, as it were, in a simple post-modern sense.[57] It has a postmodern style, first, because, as I have demon-strated, the transcendent society is a pastiche of previous conceptions of the ideal society and features these as important parts of its own structure. Sec-ond, the transcendent society as a utopian image is critical of some modern values (e.g., competition, materialism) while coopting and promoting others (e.g., humanism, spiritualism).[58] Finally, because of its eclecticism, the tran-scendent society as utopia is able to reconcile criticism and paradoxes that often confronted other utopias. Other modern utopias experienced tensions between rampant individualism (e.g., benevolent despotism, divine rule) and mindless collectivism (e.g., Orwellian totalitarianism). Cultural development often occurred together with its destructive consequences for nature. How-ever, the transcendent society is a utopian society where social control is tempered and tamed by individualism so that Big Brother does not convert one person's utopia into another's dystopic nightmare. Work and cultural development help transform people while the world of flowers and brooks remains Eden-like, maintained and protected in some mysterious way from the usual ravages of damage and exploitation.

The images of the transcendent society in the NDE stimulate a sense of interconnectedness in NDErs and those who read or hear their stories. These images appear to overcome the contradictions and problems associated with the worlds of spirit, culture, and nature. The tensions between culture and nature are reconciled in the arcadian images of people appreciating the effect of their own needs on each other and on the environment. The value of restraint is learned in the context of a new appreciation of the intercon-nectedness of human action within the social and physical universe. The many attempts to explain the NDE as a human experience have also seen several attempts to bridge the perceived gap and tension between religious and scientific paradigms.[59]

This is a special type of utopia, essentially utopian in a modern sense but at the same time featuring social elements from many ideal societies favored and pursued in the past. Because of these features, the transcendent society is able to field common criticisms of utopias by overcoming contradictions that have plagued other conceptions. Little wonder that the transcendent society and the NDE in general have captured the popular imagination. In this important way, NDEs have reawakened the pursuit of the ideal society after nearly a century of collective pessimism. This historical characteristic makes the transcendent society unique in another way.

As utopia, the transcendent society represents a special type of ideal society emerging, or reemerging, as it does, in the late twentieth century. This is because, as Kumar observed, this century has been host to the claim that "utopia is dead."[60] Too many events this century have dampened optimism and discouraged utopia. Two world wars, Nazism, Hiroshima, Stalinism, Pol Pot, the Cold War and the arms race, and recently the collapse of Eastern bloc Communism, have battered and dismantled the romantic visions of nineteenth-century utopian writers.

Despite this, some utopias have survived, for example, the kibbutzim, science fiction utopias, and New Age writing. Among other developments, the writings of Marshall McLuhan, Timothy Leary, Herbert Marcuse, Charles Reich, and Ivan Illich have all served the sporadic and apparently indomitable pursuit of the ideal society. Within the darkened recesses of twentieth-century pessimism these experiments and writings have supplied, or have attempted to supply, new ways of resolving the problems of living with the urban industrial cultures of modernity.

Kumar argued that these works have not become the central symbols for society but rather have flourished as specific visions for specific groups.[61] Certain cults, communes, social movements, and types of social theory have been peddlers and adherents of these various utopian inspirations.

Here, however, Kumar overstated his case, confusing the lack of popularity with the problem of cultural representativeness. It is true that many of the utopian visions of this century have been group specific in their attraction. However, the pursuit of alternative utopian visions as a generalized pursuit in itself has in fact been widely representative, but in a special sense. Although twentieth-century utopias have often not been representative in the content of their social ideas, they have been in the sense of their creative source. There has been widespread dissatisfaction with modern social conditions and values, and consequently a pursuit of better ones. Products of this discontent can be seen in the steady growth and acceptance of feminist, environmentalist, self-sufficiency, prodemocracy, and social network ideas, and their impact in the spheres of politics, the workplace, and the family and household, to name only a few. In this respect, utopias as forms of alternative social knowledge actually depend for their very appearance on *widespread* dissatisfaction with the existing world.[62] This complements Zaleski's observation that NDEs occur most when cultures cause social and moral dislocation and a widespread need for orientation develops.[63] So the revision that must apply to the received twentieth-century wisdom that "utopia is dead" is simply that lately in this century, a widely attractive utopia has been difficult to discern. However, the widespread *desire for and pursuit* of utopian social ideas are alive and well and historically accounted for.

In this context, the transcendent society is an exceptional utopia, not because its images and values draw from this same source of social discontent but, more remarkably, because these images do not arise from any one social group. Furthermore, despite some cultural variation in NDE imagery, the basic organization and ideas of the transcendent society remain fairly stable. Village huts that float above the ground may indeed replace cities in some versions of the transcendent society,[64] but the values of order, cooperation, kindness, and learning appear to be stable, or at least widely reported, ideas.

As such, the transcendent society as utopia provides a set of ideas widely representative in aspiration, inspiration, and constitution. It is therefore widely appealing as an ideal form of society. Furthermore, the rise and popularity of NDE imagery in industrial societies suggests and highlights a general dissatisfaction with the depersonalizing and alienating conditions within them. For those who have not experienced an NDE, this imagery becomes a rich thought exercise[65] or "mode of visualizing"[66] that fires the social imagination. To place this observation within the utopian discourse is to say that it is the conceivability of the ideas and values, rather than the achievability of any actual social system, that becomes important.[67] The traditional debate over the realizability of utopias is less important here. As Peter Beilharz observed, one of the social functions of utopian social ideas is to sharpen our understanding of current political and moral dilemmas.[68] It is this function that may characterize and take precedence in a postmodern form of utopia such as the transcendent society.

conclusion

Zaleski argued that NDEs are not widely attractive utopias. Rehearsing an argument similar to that of Kumar, Zaleski believed this is because NDEs are unable to be a widely shared basis for a new philosophy. This, in turn, is due to a lack of symbolic power that wider traditions such as medieval NDEs once had. But that perspective underestimates the attractiveness of NDEs to wider social movements, movements stimulated by rapid and disruptive social changes of the ongoing Industrial and Postindustrial revolutions. These changes have continued to fragment and disorient society through two world wars, as well as innumerable domestic and international conflicts and divisions. In this context, many people have overlooked the possibility that the attracting power of the NDE may indeed be part of a wider tradition: the pursuit of the ideal society.

A long tradition, in evidence in national politics and religion, social theory and social movements, the pursuit of the ideal society bloomed in the romantic climate of nineteenth-century Western idealism. The twentieth century has seen utopia as a social idea and an experiment falter as people exchanged hope of harmony for peace and then hope of peace for mere tolerance. But the obstacle to a widely attractive set of utopian images was also, ironically, the driving incentive for the continuation of its pursuit, albeit in small but important social experiments, in less popular but no less influential social theory and literature. The modern challenge confronting utopias has been the development of a set of images that might cross different social groups and boundaries, inspiring and uniting them in similar ways.

The transcendent society appears to be a utopia whose features can be seen as attractive to an array of different groups. As I have argued, this may be one final reason for the popularity of NDEs. Despite this popularity, the transcendent society may not be a high-profile utopia, offering as it does only a pocketful of assorted simple values and very little in the way of social programs and policies. To our recently dark notions of death, and in our pursuit of the ideal society against the even darker cynicism of our times, the ideas of the transcendent society may appear only as dim candlelight. But in beginning the long task of rebuilding optimism and a shared view of a better society, that may just be enough.

THE ACADEMIC
REACTION

The Politics of Dismissal and Apathy

7

THE RHETORIC OF
NEUROSCIENCE

There have been two principal academic reactions to the NDE. One of these has been to adapt neurophysiological concepts and theories to the phenomenology of the NDE. From these sources arise explanations of NDEs as types of hallucination. We will examine this academic reaction in this chapter. The other academic reaction, less common but equally serious, is to link these images to a general theory of psychoanalytical defense. The physiological reactions are themselves responses to the deepest fear of the human animal—the fear of death. We will discuss this academic reaction in the next chapter.

The belief in the value neutrality of scientific explanations has led, and continues to lead, to uncritical and unreflective analyses of the NDE. This is nowhere better seen than in the theories of medicine and psychology. Researchers in these areas are now turning to developments in neuroscience for their explanations of the NDE.

These are two problems here. An interest in physiology, if not balanced by an equal interest in cultural factors, can lead to reductionist understand-

ings of organic processes and responses. We will discuss the decontextualizing implications of this interest in Chapter 9. The second problem with
this interest, which partly emerges from the first, is that some neuroscience
workers are tempted to weigh into debates outside of their area of expertise.
This they do under the misleading cloak of value neutrality. However, far
from being disinterested, much of this neuroscience writing has a political
dimension. And furthermore, this is more covert than the writing of psychoanalysis because declarations of allegiance to science are more vehement
and the language of the writing is more obscure. The aim of this chapter is
to demonstrate this politics, which is inherent in many neurophysiological
explanations of the NDE.

I will argue that, despite the claim of value neutrality, more than a few
current neuroscience explanations are partisan ones. This can be demonstrated through a close inspection of their choice of language, rhetoric, and
metaphors.

In support of this argument, I organize this chapter in the following way.
First, I will provide a brief summary of the central neurophysiological explanations that currently dominate the medical and psychological discourse.
I will then examine the language, rhetoric, and metaphors that characterize
so many of these explanations. Finally, I will connect the rhetorical features
of this writing to the sociological circumstances of their rise and use: the
resistance of conservative elements in the scientific community to recent
developments in postmodernity and science; the ongoing conflict of science
with religion; and the historical rise of the medical profession and the political authority of its theories.

Neuroscientific Explanations of the NDE

One of the first medical explanations of the NDE focused on the role of
drugs.[1] Of particular interest were tranquilizers, anaesthetics, pain-killing
agents including narcotics, and stimulants such as adrenalin and ampheta-
mines. Much of the impact of this initial theory was lost when increasing
numbers of people who were not under these influences still reported the
NDE.

Today the major medical and psychological explanation appears to be a
combination of neurophysiological hypotheses that center on stimulation of
the area of the brain called the *temporal lobe*. This is sometimes referred to
as *limbic lobe syndrome*. The temporal lobe is known to be responsible for

some of the phenomenology of epilepsy. Intense emotions, involuntary memory recall, and a sense of "depersonalization"—a feeling of detachment from one's self and/or body—are frequently reported by those who suffer from abnormal disturbances of the temporal lobe. The mechanisms hypothesized to be responsible for this in nonepileptic populations are either oxygen deprivation of the brain or endorphins. In the case of oxygen deprivation of the brain, the chemicals released during these moments of physiological stress are thought to act as toxins/poisons, which in turn stimulate the temporal lobe to produce the relevant "psychosis."

Like the drug-induced explanations of NDE, the cerebral anoxia explanation is seen as only one possible explanation. This is because it is clear that not all NDEs occur to those who are unconscious and near death. As we have seen with shipwrecked castaways, most of the characteristics of NDEs can involve people who are not dying, but rather who are in good health but in physical danger. The occurrence of NDEs in people who are not actually dying has been widely recognized.[2]

When oxygen deprivation does not appear to be an appropriate hypothesis, a theory of endorphin action is propounded. Endorphins are morphine-like substances produced by the brain during periods of high stress—for example, at times of athletic exertion or trauma. These chemicals are thought to be responsible for the failure to experience some types of pain (e.g., from knives or bullets) or for the ability to endure what would normally be considered disabling pain (e.g., in long-distance running).

Although endorphins are occasionally thought to do more than produce analgesia,[3] it is rare to observe runners experiencing life review, tunnel sensation, and meetings with deceased relatives. It is also difficult to equate the presumably diverse physiological states of shipwrecked castaways, meditators, LSD users, and the dying with those of long-distance runners.

Given this methodological ambiguity, a psychophysiological theory of hallucinations has been developed. Sometimes this is accompanied by the previous hypotheses of anoxia and endorphins[4] and sometimes without these supports.[5] In this explanation, the NDE is due to a brain that is starved for external stimuli. These are hallucinations due to sensory deprivation. Because the brain is dying or because sensory input from the outside world is compromised, the central nervous system begins to feed off its own images. This inward pattern of thinking can be made relevant to experiences of meditation, LSD use, and social and physical isolation. Such internally produced images come from two sources: the visual cortex and the malfunctioning temporal lobe.

From the visual cortex, optical neurons fire randomly to produce what Kluver, among others, calls *form constants*.[6] These are lattice, cobweb, tunnel, and spiral images that dominate most, if not all, hallucinatory experiences. Then, with further changes in the central nervous system, the brain turns to its memories for images of life review and reunion. For the moment, we must leave aside the anthropological evidence suggesting that the tunnel sensation and life review are not cross-cultural and hence not amenable to universal theories of physiology (Chapter 2).

However, it is important to note that many of these explanations are offered in the spirit of inquiry into the medical mechanics of the NDE. In this context, these theoretical exercises neither challenge, nor add to, nor attempt to dismiss the transcendental claims or philosophical meanings of the NDE.

For example, Neppe remarks:

> The anatomical or physiological link between NDEs and any pattern of brain functioning does not in any way imply a direct etiological link. NDEs may reflect genuine patterns of functioning outside the brain, modulated by a particular psychophysiological mechanism. Alternatively, NDEs may be the epiphenomena of an experience purely within the brain, without any outside reality playing a role.[7]

This echoed the earlier sentiments of Carr:

> To elucidate previously unstudied reflexes of the moribund state would not of course reveal deaths meaning any more than elucidating perinatal reflexes reveals the meaning of life . . .[8]

A recent article by the French physician Jean-Pierre Jourdan has attempted to examine the diversity of perceptual experiences in mystical states, including the NDE, and to link these to neurophysiological correlates. But his penetrating theoretical exercise involved exploring mechanisms, not arbitrating among conflicting epistemologies:

> The deeply spiritual aspect of these experiences eludes for the moment any objective research, but it exists nonetheless. I hope I have shown that it is possible to draw a bridge, however frail, between science and transcendence, and that spiritual matters can be studied without necessarily making the opposing mistakes of either a scientific reductionism or a blind mysticism, which may be but differing appearances of the same lack of curiosity.[9]

But not all such medical theorizing has displayed such impartiality and philosophical reserve. Many other theorists, as Zaleski observed, overstate their case, "treating as cause what is at best analogy."[10] The writings of these theorists reveal a political rhetoric that, intentionally or not, places their language of explanation firmly in the tradition of medical reductionism, professional dominance, and open conflict and rivalry with religious ideas. We need only examine some of this rhetoric to see how their medical explanations are part of this wider political and historical agenda.

Language

When the science philosopher Bulhof began reading scientific texts, she held the rather common view that they would be written in a dispassionate, abstract style.[11] Their supposedly objective, descriptive narrative style would be as interesting as the chemical analysis on the back of a soap box. However, the more scientific papers Bulhof read, the more she became aware that "expressions of feelings, questions, exhortations and exclamations, rousing of sentiments or flights of fancy" were common. Indeed, "all writing has a rhetorical side" and scientific work, popular views to the contrary, is no exception. In this respect, it is not the actual content that is at issue here but rather the rhetorical devices used to "sell" the content—the devices of persuasion incorporated in the linguistic style.

Mills understood this when he argued that language is a *social* device that carries with it implicit social evaluations and exhortations.[12] But more than this, writing, all writing, is action that is influenced by others in two important ways. First, the author's view of the potential readers influences the writing. In this respect, the successful writer of a scientific paper, apart from having an original or novel idea/data, must also successfully anticipate and then negotiate the questions, prejudices, and objections of his or her peers.

Second, Mills, paraphrasing the anthropologist Malinowski, argues that thinking "influences language very little" but that language decidedly shapes thinking. This shaping occurs through the limited nature of the meanings of the words or expressions themselves. Language is shared meaning, with revisions coming only after explicit negotiation; otherwise, miscommunication is the result. In this sense, writing is a rhetorical exercise designed to persuade and lead the reader to the writer's world of thought and feeling. A study of NDEs is no exception, and as Seigal reminds us, such a study "is highly dependent on the words, pictures, and other symbols used in description."[13] On Seigal's advice, then, let us begin an inspection of the words and expressions in medical and psychobiological works on the NDE, beginning with Seigal's own work.

In his opening remarks concerning the origins of belief in the afterlife, Seigal writes:

> The most *logical* guess is that consciousness shares the same fate as that of the corpse. Surprisingly, this *commonsense* view is not the prevalent one.[14]

Note the Seigal's materialist view here is identified with the voice of reason; other views are, by implication, irrational. He presents his view as being obvious, a practical one unclouded by false hopes or illogical images. Nevertheless, he also observes that despite its "common sense," it is not commonly held, a rather unfortunate choice of word given the cultural context. However, the implication is clear: If the reader has a view that diverges from Seigal's materialist view, then, by implication, that view may be *illogical* and *lacking in common sense.*

Expressions meant to entertain but also derisory to authors whose views differ from Seigal's are also part of Seigal's rhetoric. For example, a summary of Koestler's theory of the afterlife, which attempts to link dualism with recent notions in physics and biology, is followed by the inexplicable comment from the movie *Star Wars,* "May the Force be with you, Arthur!"[15] This comment is not designed to encourage readers to take alternative ideas about death, other than Seigal's own, seriously.

Koestler's treatment is not an isolated example. Ian Stevenson, a psychiatrist well known for his work with children who claim to remember past lives, is described as a "ghost chaser."[16] Never mind that Stevenson's work has been taken seriously by some of the world's most respected medical journals. Seigal encourages readers to see this as a laughing matter.

Other language employed by Seigal equally demonstrates an enthusiasm for loading his terms.[17] His dedication to a singularly empiricist view of the world does not entertain other epistemological possibilities. Sincere reports of unusual experiences by other writers are described thus:

> . . . the common features of their descriptions are viewed as indicative of a common *objective* reality—never a common *subjective* reality.[18]

Here reality is presented as clearly understood and divided between the high-status (objective) reality and the less credible (subjective) one. The epistemological difficulties in separating objects and experiences from their phenomenological, cultural origins and attributes are presented as entirely unproblematic. The issue can apparently be dealt with using this form of

binary analysis, philosophical, psychological, and anthropological reservations and complexities notwithstanding. Seigal is not alone in this subtle use
of language to load "scientific reasoning" his way.

Blackmore, another psychologist who looks to a neuroscientific model for
her theory of the NDE, is also adept at this style. Readers who find Blackmore's model of consciousness difficult to relate to are told that former models that use a central sense of self are probably

> an *illusion* we all live in [that is] that there is someone lurking inside who
> makes the decisions, is the "center of consciousness."

We are told that it is

> *progress* in *modern neuroscience* which is undermining the idea of any central
> self . . . [19]

Three implications emerge from this rhetorical choice of words. First, if we
do not accept Blackmore's position and instead hold fast to some essentialist
idea of the self, then we are deluded about consciousness. And, as if this
error is not shameful enough, we are reminded that it is not only an illusion
but an outdated and backward one as well. To support this charge, neuroscience is portrayed as a unified science with little or no dissent and little or
no diversity of theoretical opinion. This astonishing achievement has apparently escaped the attention of people as diverse as Karl Popper,[20] Thomas
Kuhn,[21] Michel Foucault,[22] and Ludwig Fleck,[23] to name only a few of the
science historians and philosophers who have documented the opposite in
science. Furthermore, changes in the opinion of mainstream nueroscientists
are described as "progress" rather than simply the current view, a view that
may be vindicated or rejected next year.

We can see that Seigal's and Blackmore's language is designed not simply
to present a psychobiological theory of hallucinations but also, in the process,
to appeal in loaded terms to fellow believers of this theory and to question
the *rationality* and *credibility* of other views and their advocates. However,
as I will demonstrate, Seigal and Blackmore are not alone.

Other rhetorical examples of loaded-language use can be seen in the work
of Saavedra-Aguila and Gomez-Jeria.[24] The *Journal of Near-Death Studies*
dedicated a special issue to their neurophysiological theory of the NDE, and
this too, despite its declared allegiance to scientific neutrality, is replete with
rhetoric. Much of their work uses descriptors designed to give the NDE an
abnormal medical appearance. The NDE is described as part of temporal
lobe "dysfunction"; as a neurotransmitter "imbalance";[25] and as an example

of "abnormal brain functioning";[26] the model they put forward is a "patho-physiological" one;[27]

> The life review in the NDE can be understood as an *abnormal* retrieval of episodic memory contents by the *dysfunctional* limbic areas.[28]

Noyes, in commenting on Saavedra-Aguila and Gomez-Jeria's paper, notes this emphasis on the "abnormality" of the NDE and asks why the implications of this idea are not discussed.[29] A neurophysiological model is not necessarily abnormal unless the phenomenon in question has been demonstrated to be abnormal. In this case, the writers either assume this to be so or, alternatively, have drawn so much of their model from aberrant neurophysiological examples that the proposed model exhibits these origins as a kind of theoretical birthmark. The NDE model proposed seems abnormal because it is, in reality, a collage of borrowings from pathophysiology, a branch of physiology dedicated to the study of the diseased and abnormal physical processes of biology.

In addition to the judicious and careful use of words to shape an impression and lend support to a theory, rhetoric is also enlisted by many medical and psychological writers on the NDE.

Rhetoric

Rhetoric may be described as a persuasive strategy using comments that are extraneous to the line of argument or evidence offered. Consider the following example:

> The use of the scientific method to construct hypotheses constrains the verbal system so as to exclude personal beliefs and superstitions.[30]

And a similar piece of optimism from Blackmore, who states:

> It is part of our normal use of language that if there is joy there has to be someone experiencing it. I do not think it is part of *fact.*[31]

Both of these writers ignore the integrated nature of thought and observation.[32] Facts are not obvious or understood separate from the theory that supports them.[33] In any case, reliable hypotheses, however well constructed, and facts, however widely agreed on, merely temper personal belief; they do not exclude it. The contemporary state of psychiatric theories of schizophrenia[34] or physiological theories of multiple sclerosis[35] are evidence that facts rarely settle debates, let alone constrain personal beliefs.

Persinger provides another example of rhetoric, although not as obvious as the above illustrations.[36] This is because the wording is subtle, coming as it does in the middle of a technical clinical discussion of the possible long-term sequelae of an NDE.

> The type of personality change should be specific to the "temporal lobe pattern" that is seen clinically; symptoms include increasing circumstantiality, religiosity, philosophical pursuits, widening affect, and a sense of the personal (Bear & Fedio, 1977)[37]

First, note that personality changes are written into the narrative as "symptoms," a gratuitous and medicalized ascription of pathology and deviance that may be neither. Second, religiosity and philosophic pursuits are associated with epileptiform activities, though, of course, this is merely a report of the Bear and Fedio (1977) study. But when we turn to this study, we discover that Bear and Fedio do not themselves refer to these personality changes as "symptoms" but rather by the more behaviorally neutral term "traits." They argue that these traits are *not, in themselves,* indicative of psychopathology and do not "correspond to any standard psychiatric descriptor."[38] Why, one might ask, does Persinger refer to these personality changes as symptoms if they are neither abnormal neurological nor psychiatric indicators?

The third issue to emerge from these remarks is that Persinger selectively mentions circumstantiality, religious and philosophic interests, widening of affect, and a sense of the personal. But Bear and Fedio stress the importance of the appearance of *all* of their identified traits *in concert.* Obsessionalism, humorlessness, dependency are cited, and when Bear and Fedio mention "affect," they single out anger, emotionality, and sadness. These traits are widely recognized *not* to be associated with the NDE. These traits, which would shed critical light on their relevance to the NDE, are conveniently not mentioned by Persinger.

As we have seen with Persinger, the use of rhetoric can also lead to overstatement of one's case. I will provide three further examples. Saavedra-Aguila and Gomez-Jeria state:

> Recent neurological analysis of some religious events, such as visionary experiences from written medieval sources (Kroll & Bachrach, 1982), and St. Paul's

ecstatic visions (Landsborough, 1987) and sacred painting (Janz, 1987), which seem to correlate well with epileptic phenomenology, suggest that we are on the right path in separating physical elements from metaphysical ones.[39]

In regard to this remark, it should be noted that Kroll and Bachrach[40] do not provide a "neurological analysis" but rather a psychiatric one that concludes that over one-half of medieval visions cannot be attributed to abnormal states. Landsborough's analysis of St. Paul's ecstatic vision contains the following qualifiers:

We do not know whether Paul showed any abnormal physical signs

and furthermore,

Nothing is known about Paul's past health and family history.[41]

Landsborough's analysis is based on a simple content analysis of the visions and on his own biblical interpretation of stories about Paul. This is not a neurological analysis unless hermeneutics has become a branch of medicine.

When we turn to Janz's analysis we discover, once again, a semiotic exercise interpreting the significance of a painting that juxtaposes Christ's transfiguration scene with that of a scene in which a boy undergoes an epileptic seizure.[42] The epileptic features are not in question but rather the significance of the dual images to an appreciation of the art.

Saavedra-Aguila and Gomez-Jeria exhibit more confidence and impute more specificity to these sources than do the sources themselves. Indeed, it is a pity that Saavedra-Aguila and Gomez-Jeria, in mentioning Kroll and Bachrach, do not cite the rather pertinent caution offered by those authors:

There is a tendency to over-explain or over-diagnose all or most visionary experiences, since we live in a time when altered states of consciousness and transcendental experiences are suspect.[43]

Roberts and Owen argue, in concert with others they cite, that NDEs cannot be considered death experiences because death means ipso facto nonrevivable.[44] NDErs have not "returned" from death because they have not died, an argument also presented by the Catholic theologian Hans Kung.[45] But as Zaleski rightly points out:

Whether near-death experience occurs in the grip of death or only in the face of death, it may still constitute a revelatory encounter with death.[46]

Finally, Seigal argues that, after all, hallucinations can occur in conscious people "when triggered by emotional states surrounding death."[47] He cites Matchett for this remark, but Matchett does not offer any evidence that hallucinations are the actual operative mechanisms.[48] Matchett merely characterizes the sightings of deceased spouses by their widows as hallucinations; this is a phenomenological analysis rather than a neurophysiological and clinical evaluation. Matchett is concerned with the issue of grief resolution, not hallucinations for their own sake. I now turn to the rhetorical use of metaphor by advocates of neuroscience.

Metaphor

A metaphor is a figurative representation based on the principle of resemblance. It stands for a set of processes or circumstances and is occasionally identified with those processes. Metaphors, then, are both a persuasive and a didactic strategy in many narratives.

Jewson argues that metaphors are integral to medical cosmologies, that is, they are essential to all paradigms or models of scientific understanding.[49] In this way, through the employment of metaphors, the scientific world is able to see in certain prestructured ways. This allows conjecture and theories based on these metaphors to be tested, of course, but in this process such models also create ways of not seeing.[50] In neuroscience there has been a succession of models that have provided hypothetical analogies for the human brain. Each of these models has evidence and advocates. Among the major metaphors are the telephone cable, the computer, and the holographic model.[51] Few scientific advocates of any of these models mistake them for reality. They understand the place of metaphor in the development and debate of ideas within the scientific discourse. Dennet, for example, whose best-selling book *Consciousness Explained,* is the basis of some NDE researchers' model of the brain, accepts this point.

I haven't replaced a metaphorical theory, the Cartesian Theatre, with a non-metaphorical ("literal, scientific") theory. All I have done, really, is to replace one family of metaphors and images with another, trading in the Theatre, the Witness, the Central Meaner, the Figment, for Software, Virtual Machines,

Multiple Drafts, a Pandemonium of Homunculi. It's just a war of metaphors, you say—but metaphors are not "just" metaphors; metaphors are the tools of thought.[52]

I will now examine three metaphorical devices that purport to explain part or all of the NDE. One of the metaphors favored by Seigal shows how normal memories are usually suppressed by the same action in the brain that permits outside information from entering.[53] If, however, that input decreases, for whatever reason, memories and images are released and "dynamically organized," to appear as "hallucination, dreams or fantasies." In making his appeal for the credibility of this idea, Seigal does not provide experimental data but rather summarizes a metaphor used by the psychiatrist L.J. West.

> Picture a man [sic] in his living room, standing at a closed window opposite the fireplace and looking out at the sunset. He is absorbed by the view of the outside world and does not visualize the interior of the room. As darkness falls outside, however, the images of the objects in the room behind him can be seen reflected in the window. With the deepening darkness the fire in the fireplace illuminates the room, and the man now sees a vivid reflection of the room, which *appears to be outside the window.*[54]

Up to the last few words, the metaphor is an attractive one. However, unless the man described in the room is himself hallucinating, he would not mistake the internal scene for one appearing outside his window. The images *on* his window would look exactly like that—images on the surface of the window. Both the clarity, or lack of it, and the contents of the competing images would and do appear different. This is entirely the case with dreams or fantasies, which rarely appear to be part of the outside world. On the other hand, hallucinations and hypnagogic states may appear to be part of the outside world, but the analogy presented above is not a successful metaphor for them. And since both hallucinations and dreams enjoy a great variety of imagery, we are still no closer to a metaphor from Seigal that can model the less diverse imagery of the NDE. So, the metaphor neatly describes dreams and fantasies. But considering that Seigal's intention is to conflate dreams with hallucinations, and combine both of these with NDEs, the metaphor is technically poor and reductionist.

Toward the end of this same work, Seigal employs another metaphor to explain how his theory compares with religious explanations of "life after death experiences."

In the past, dying and death were often accompanied by fear and loneliness, as if the individual were possessed by Pan, the Greek god of lonely places and panic.

But if we accept Seigal's view of these experiences as hallucinations,

we can counsel the dying to take the voyage not with Pan at their side, but with Athena, the Greek goddess of wisdom.[55]

There it is, then. The choice is either to join the hysterical throngs who wish for an afterlife or the wise ranks of those who know a hallucination when they see one. Despite the clear choice offered by Seigal, the decision presented to the reader remains rather difficult. This is because the options are cast in such narrow and conceited terms.

Blackmore and Troscianko also provide metaphors for our consideration but, once again, they confuse the representation with its object.

The light at the end of the tunnel is induced by randomly firing neurons. It is not just imagination. It has a *very definite* physiological origin.[56]

Here Blackmore and Troscianko draw a parallel and then plead for its acceptance as reality. But I can also equate tunnel sensation with traveling through real tunnels. This too is a parallel, has a genuine correspondence with the NDE tunnel experience, and, incidentally, is commonly used by spiritualists and New Age theorists. However, the evidence for the reality of either is poor, both merely spending their respective energies on developing the details of their metaphors. As even Blackmore and Troscianko admit, simulation is the best they can do since "obviously, we cannot open up someone's cortex and apply the hypothesized stimulation that way."[57] In the meantime, the parallels conveyed by the respective metaphors are used to appeal to and confirm the a priori beliefs or assumptions of their respective audiences.

The Historical and Political Context of Neuroscience

There are several likely historical and political reasons that prompt certain sections of the neuroscience community to embellish, with such forceful rhetoric, their otherwise worthwhile work. As we have seen, the rhetorical appeals in the above examples seem to go beyond the usual tempered styles of persuasion that one is accustomed to in the hard sciences. The extreme rhetoric that emanates from some sections of neuroscience can be seen as a reaction to public criticism of the political authority of medical explanations. This threat comes from several historical and contemporary sources.

MODERN VS. POSTMODERN KNOWLEDGE

First, in a general way, some of the rhetoric may be seen as the conservative reaction of some of its members to recent social changes, particularly those in postmodernity and science. The postmodern model of science, according to Turner, is one in which a single overarching and comprehensive idea of rationality has broken down into several competing and conflicting ones.[58] Scientific rationalism, for example, is often now overtaken by or made subservient to economic rationalism.

Furthermore, the paradigms (science, reason, the Enlightenment, and humanism) that emerged from a critique of religion and assumed the form of utopian ideals in this century have fallen short of their collective promise. This has led, as I have argued in the previous chapter, to a search by both the lay and academic publics for alternative paradigms of knowledge and morality.

This undermining of the great unifying discourses of the nineteenth and twentieth centuries has not gone unnoticed, or undefended, by those whose livelihood and interests are most closely identified with it. Skepticism, as a social movement that vigorously defends the primacy of science, reason, and humanism[59] has fought bitterly and publicly to defend the honor of these historical traditions. The skeptics, more than any other social or political group, have encouraged the use of disparagement and reductionism as part of their active campaign of aggressive debunking and rhetoric.

Less noticed in all this agitation is the fact that the skeptics' responses are a sectional reaction to disillusionment by lay publics to the very narratives that skeptics wish to champion. In all of the publicized commotion associated with skepticism, it is important to remember that the skeptics are not attacking but rather defending. They do not wish a return to the dominance of religious ideas, ideas that are not amenable to testing or questioning. In that context, the skeptics offer only answers that survive or thrive under test conditions, or conditions within the rational framework of logical positivism. This is a position shared by some of the neuroscience authors—for example, Blackmore (incidentally, a member of a skeptics' organization).

SCIENCE VS. RELIGION

This ongoing historical dispute that science has experienced, but continues to direct toward religion, forms another important backdrop to the enthusiastic rhetoric of neuroscience. Because of the suggestion that the NDE may be a glimpse of life after death, skeptics and humanists within medicine and psychology have once again taken up the cudgels against this latest round of "religious" ideas. Note that all the authors from whom examples of rhetoric are taken cast religion and religious ideas in terms of rival explanations.

Blackmore and Troscianko devote part of their discussion to "occult" explanations of tunnels in the course of their review of the subject.[60] Roberts and Owen also deal with "spiritual theories" in their review of causes of the NDE.[61] Seigal couches his entire paper in terms of the "scientific" alternatives to religious theories of the afterlife.[62] Finally, Saavedra-Aguila and Gomez-Jeria remark that previous attempts to examine the NDE appear "to consider models rooted in previous beliefs (for example, astral body explanations)."[63] They go on to assert that these theories appear to have little support given "present knowledge."

NDE theorists from medicine and psychology continually display a compulsion to address their remarks not only to their own peers but also to those in the clergy. And despite the little credence attributed to religious ideas by most of these theorists, their regularly occurring remarks demonstrate a strange preoccupation with them—a preoccupation that has a long tradition.

Leary notes that the New Psychology that pushed for physiological explanations early in this century was born from a critique of religious and psychic research ideas about the person.[64] But both medicine and psychology contributed to, and developed during, the decline of religion or religious credibility during the Enlightenment. At that time, medicine's discourse with religion moved beyond the informality and complicity it had enjoyed with the ruling classes during the Middle Ages.[65] In psychiatry, medical people began to object to the burning of witches, while leaders in public health began to decry the sanitary, work, and housing conditions of town populations.[66] But medicine's and psychology's twentieth-century involvement with society extended considerably beyond social criticism of living conditions.

The development of psychiatry, in both its neuropsychiatric and its psychodynamic aspects, has attempted to claim authority over knowledge about the self. Psychology has laid some claim to that authority, too, in its educational, industrial and clinical aspects.[67] Medical and psychological ideas have sought to compete directly with religious ones. Medicine in particular has been so successful in this competition that Zola was moved to remark:

today the prestige of *any* proposal is immensely enhanced, if not justified, when it is expressed in the idiom of medical science.[68]

This echoed and extended Polanyi's lament:

> In the days when an idea could be silenced by showing that it was contrary to religion, theology was the greatest single source of fallacies. Today, when any human thought can be discredited by branding it as unscientific, the power previously exercised by theology has passed over to science; hence science has become in its turn the greatest single source of error.[69]

The status of medicine has so increased over the last century that its authority over every major aspect of self and lifestyle has usurped the previous political incumbent—religion.[70] This is not the place to explore in detail the political and sociological reasons for this medical dominance. It is sufficient simply to note that, for most of its ideas and practices, medical science has enjoyed much community support.

Nevertheless, this support has not always been and is not always so unqualified. A critical community discourse does exist, and the emergence of the New Age movement has benefited from this by becoming an explanatory competitor. This is the final and most significant way in which some sections of the neuroscience community feel threatened. To illustrate this rather complex point, I will present an example of a positive discourse for medicine. I will then compare this with the less appealing attempt imposed on the NDE.

NEUROSCIENCE VS. THE NEW AGE

The example of AIDS research is instructive in this regard and provides a good comparison with neuroscience research into the NDE. White and Willis describe how the discourse on AIDS is typically an affirming one for medical science.[71] The problem at the center of this discourse is believed to be a virus. This has historical similarities to the epidemiological problem of bacterial infections, a problem largely controlled by public health programs and rising living standards in this century.[72] Nevertheless, in the popular imagination, it was the development of antibiotics that is seen as instrumental in combating the major bacterial diseases of previous times. This apocryphal imagery fuels community views of the medical "fight" against AIDS and is an important factor in the support medical science enjoys from the community. As White and Willis remark, the community sees AIDS as a virus which we do not understand, but with the progress of science we soon will.[73]

In the middle of this relationship between medical science and the lay public are some sections of the social science and gay activist communities. These groups point out the value-laden nature of science. They emphasize the importance of cultural factors in determining what counts as real knowledge and the moral meanings ascribed to diseases. Of concern also are the ways in which diseases themselves are constructed. This involves focusing on the decisions made by medical people: what counts as a "medical problem" and whether that "problem" is seen in purely biological/interventionist terms or, for example, in social/preventive terms. Because these groups are aware that the major inroad against past infectious diseases was social rather than laboratory based, it is the social issues and processes that they emphasize.

Whatever the merits of this social science position, it is, neither understood nor seen as critical by the lay community. For the community, AIDS is an entirely medical problem to be dealt with by those who "know best" in this area—medical scientists. For medicine, then, AIDs can be seen as a positive discourse for scientists, a discourse that privileges their professional metaphors and definitions of the problem. This, however, is not the case for neuroscience studies of the NDE.

When accounts of the NDE appeared in the community, first in popular books and then through continuing media coverage, it did so in the form of "life after death" stories. These stories, as mentioned earlier, appeared as accounts that defied medical explanation. But as I argued in Chapter 5, this was only part of the general disillusionment people were experiencing with the medical community in regard to issues surrounding death and dying.

In the area of terminal illness and medical disclosure, doctors were shown to be untrustworthy. In attitudes toward death, they were seen as either rejecting, and thereby isolating the dying, or as aggressive, resuscitating patients with little or no regard for personal dignity or quality of life. The medical profession could not even show a united understanding about what seemed to be a rather simple matter for any lay person: the definition of death.

So, in that context, when medical science began to pronounce on the dramatic images of the NDE, it did so at a low point in its credibility on these matters. Psychology, its interest in these matters drawn from neuroscience too, was greeted with the same cool response from the lay establishment. Unfortunately, this tended to fuel rather than moderate the rhetorical appeals of neuroscientists, who were only too aware of the current climate of suspicion and criticism.

Between medical science and the lay community, in this instance, there was no effective social science response. There was, however, a large and growing New Age movement that was easily able to accommodate the demands of the community for an explanation of the NDE.[74] But the role of the New Age movement was not, I believe, to act as the latest repository for hopes of life after death. In the previous chapter, I argued that broader, more politically based ideas might be responsible for the attraction to NDE imagery. However, complementing this desire for inspirational ideologies is a social vacuum created by the deficiencies of medicine itself. This is an area of professional neglect readily filled by the person-oriented "alternative" occupations and ideas of the New Age. Another look at medical history will explain this situation further.

In his social history of medicine from 1770 to 1870, Jewson documents the development of modern cosmologies of healing.[75] Up to 1770, doctors were dependent on their patrons for work. In that context, patients were powerful. It was incumbent on doctors to produce an explanation of illness that accorded with the patient's experience. At this time, organic systems were primitive in development and did not play a major role in the diagnosis and treatment of disorder. Rather, the patient's subjective account *of his or her experience of illness* was the basis for theories of sickness. Jewson describes this cosmology as "Bedside Medicine":

> Thus in effect the patient appeared in the cosmology of Bedside Medicine in a guise similar to that in which he appeared to himself, i.e. as an individual and indivisible entity.[76]

With the development of hospital medicine, with its interest in anatomy and hidden internal bodily processes, doctors concentrated on patients' accounts only insofar as those accounts yielded information about symptoms. Beyond this, doctors attempted to elicit clinical signs that would yield further information of which the person had no knowledge. At this critical point in medical history, then, the patient's experience begins to be separated from the production of medical knowledge. But this development does not stop here.

Laboratory medicine takes the concern for organic events and structures to the "cellular level" and the level of "physico-chemical" processes.[77] In this process, it not only dispenses with the patient's experience but also no longer even requires his or her presence. A sample of tissue from the patient's body is all that is required for examination and theorizing. This is what Jewson describes as the disappearance of the "Sick Man" from medical cosmology, and it is this historical process that inadvertently contributes to the power of New Age explanations of the NDE.

The lay community is caught between two compelling experiences. On the one hand, they must attempt to explain what appears to be the inexplicable experience of some of its members, the NDE. On the other hand, neuroscientific explanations from doctors and psychologists, apart from being obscure and seemingly cut off from the personal experiences of NDErs and their ardent listeners, have had a poor record in matters of death and dying.

In this context, New Age explanations take as their basis the literal experience of the NDE. Attempts are then made to weave a theory around it, employing and linking similar phenomenological sources in Eastern religions and Western spiritualism. Leaving the problem of content aside for the moment, New Age theories of the NDE validate the *personal experience* even when they may not necessarily vindicate religious or materialist beliefs. These theories effectively validate personal experience because they provide reassurance of the *normality* of the NDE experience, avoiding the stigmatizing medical images of hallucinations, drug side effects, and mental illness. New Age theories emphasize the positive aspects of the experience and help people understand their NDE as part of the wider life of personal and social reality.

As for content, simple belief in an afterlife is not sufficient in performing this task of validation. If it were, the NDE would be an advertisement for organized religion, but no such association has been observed. This is because both religious and recent medical reactions to personal visions have one historical feature in common. As Kroll and Bachrach observe:

> when the visionary is a person of low status, such as a child or an unknown cleric . . . much skepticism is voiced.[78]

That inability of medicine (and religion) to affirm and to convey an understanding of common personal experiences has undermined neuroscience's ability to be taken seriously. This situation has earned New Age explanations a hearing from the lay community. In this way, returning to the analogy of the AIDS discourse, the NDE has provided a positive discourse for New Age theories but a negative or nonaffirming one for neuroscience. Like the social sciences in the AIDS discourse, neuroscientific explanations are neither understood nor seen as critical to understanding the NDE.

Ironically, neurosciences in the NDE discourse find themselves in the same political position as social sciences and gay activists in the AIDS discourse. In this position, both are seen by the lay community as hinderers

and critics of the community's preferred explanations and storytellers. As I
noted earlier, popular community preferences are not necessarily forged or
developed from the esoteric historical and epistemological debates of academics. This is more the case than ever before because, in the postmodern
electronic media, the blurring of divisions between pop and intellectual authority is increasing.[79] And while the lay public tends to accept a medicalized
discourse about AIDS, it does not yet accept that authority in matters regarding death.

This skeptical situation surrounding neuroscientific explanations has led
some of its advocates to become disillusioned and annoyed at their lack of
support. Theorizing has become more strident and the rhetoric more shrill
as many of these theorists try harder to persuade. In the service of this task,
some neuroscience writers have displayed impatience, sometimes adopting
an unfortunate reductionist and dogmatic style. That style, ironically, has
made their explanations less rather than more credible, thereby adding to
rather than subtracting from their professional troubles.

conclusion

In our examination of the rhetoric of neuroscience, we have seen that the
technical and rhetorical aspects of scholarly discussion are closely interwoven.
This is true of all writing, including the book you are currently reading. But
in scientific writings about the NDE, some authors have made a consistent
attempt to give the impression that they are above these literary and social
realities. For these authors, it has been important to reify the scientific enterprise as the embodiment of rationality and cultural neutrality. They attempt to build on the popular stereotype of themselves as scientists who are
dispassionate people concerned only with empirical evidence. The scientific
establishment, they would have us believe, is not given to the intemperate
appeals so often seen in baser pursuits such as religion or politics.

But as we have seen, science and medicine do not simply persuade with
evidence. Evidence is not enough. As Nicholson and McLaughlin note:

> the means by which competing statements about the real world are assessed is
> evidently a social one—involving not only prior beliefs and expectation but
> also power, authority and status.[80]

As I have tried to show, this statement is as true for AIDS research as it is for NDE research. Empirical evidence has not enabled the social sciences to seriously challenge the laboratory-based, interventionist medical view of AIDS. Medicines historical authority and metaphors about infectious diseases dominate the popular imagination in such powerful ways that the empirical evidence and cogent argument that might challenge them have little political strength.

In precisely the same political terms, neuroscience will not easily overcome the failures of medical authority in matters about death. The abstract, laboratory-based metaphors provide little competition for the person-oriented ones supplied by New Age writers. Unless the dual problems of credibility and relevance are addressed, modern developments in neuroscience will interest few beyond its own ranks. And if neuroscience wishes to enter the *philosophic* domain, to claim that NDEs—or smiles or dreams—are *nothing but* physical mechanisms, it will need better and more credible equipment for its purposes than mere physics, chemistry, and rhetoric.

8

PSYCHOANALYZING A RABBIT
NEAR DEATH

Stories about survival of death always attract interesting reactions from academic circles, particularly those influenced by psychoanalysis. But despite the widespread curiosity that these tales elicit, the predicability of the explanation generated must impress even the most casual observer. From a diversity of academic sources, ranging from anthropology to semiotics to psychiatry, the underlying assumptions that drive many of those explanations have a remarkable sameness. To cite a representative view, anthropologists Maurice Bloch and Jonathan Parry argue that the theme of rebirth and tales of the afterlife in the context of death "deny the finality of death."[1]

That materialist claim concerning the "finality" of death is assumed to be, and is often presented as, a culturally neutral one. Consequently, those who resist the claim or who choose to believe otherwise are viewed as "religionists" of one persuasion or another or, alternatively, as emotionally troubled. So, if religion or ignorance is not the explanation for this "deviance," a psychoanalytic theory of an unconscious defense mechanism (e.g., denial, wish fulfillment) provides another explanation.[2] The resultant effect of this style of discourse is the lasting impression of an academic community unable to free itself from its nineteenth-century reaction to religious ideas. Nowhere is this psychoanalyzing tendency more clearly evident today than in social and literary explanations of death and loss in children's stories.[3]

My task in this chapter is to provide an alternative reading of an example of children's literature that celebrates images of death incorporating ideas of renewal. I have chosen *The Velveteen Rabbit*[4] as my case example for two reasons. First, this story contains a famous portrayal of the themes of death and renewal; interestingly, the images described are strikingly similar to the NDE. Second, the academic story of *The Velveteen Rabbit*, a recent development involving attempts by some critics to substitute psychoanalysis for narrative analysis, has implications for an interpretation of NDEs that are worth closer scrutiny. The purpose of this task is to demonstrate that ideas about death that encompass personal survival are not better understood when analyzed in terms of psychoanalytic defense theory. Indeed, the habitual application of this framework to all manner of death imagery, including the NDE, becomes a reductionist practice that restricts rather than enhances our understanding of different cultural meanings of death.

The Velveteen Rabbit has attracted psychoanalysis for its refusal to acknowledge the finality of death—the Rabbit's death scene fails to result in obliteration of the rabbit's life. The fact that the Rabbit's death does not conform to the materialist view of death has made at least one critic complain that the Velveteen Rabbit's image of death is phony. This has recently prompted Steven Daniels to channel the analysis into a study of the psychodynamics of defense, motive, and symbol.[5] My view is that this approach is both unnecessary and unnecessarily decontextualizing.

The theme of renewal and survival in the face of death *is* a necessary narrative device for the support it gives to more important themes, at least for young readers. These broader themes speak to the mutual interdependence of relationships and the triumph of love in the face of change and transformation in life, particularly in the context of growing up.

In support of this argument, I organize the chapter as follows. First, I will provide a brief summary of the story of *The Velveteen Rabbit*. This will be followed by a discussion of the recent critical concern over its images of death and renewal. I will then provide an alternative reading of the images and will argue that this interpretation is consistent and more in keeping with the spirit and values of the narrative as a whole. The final part of the chapter will sketch some implications of this exercise for other interest areas within near-death studies.

The Story of The Velveteen Rabbit

The Velveteen Rabbit appeared in 1922 under the pen of Margery Williams. Accompanied by the charming lithographs of William Nicholson, it quickly became a children's story classic.[6] In 1971 it won the Lewis Carroll Shelf Award, and in the 1980s, when the copyright ran out in the United States, several new editions with different illustrations appeared.[7] At one time, its popularity was so great that one reviewer described it as having "become a cult in the United States," noting that parts of the book were regularly quoted in sermons.[8]

It is a simple story written for children between the ages of five and nine. The Velveteen Rabbit begins his life in the story as a gift inside a boy's Christmas stocking. After the initial excitement of receiving the Rabbit, the Boy becomes interested in other toys found among the chaos of the Christmas paraphernalia. For a long time after that first day the Rabbit lives forgotten, with other neglected toys, either in the toy cupboard or on the nursery floor.

He is befriended by an old Skin Horse. Deep into the long, silent nights, the Skin Horse and the Velveteen Rabbit hold long conversations. The most critical discussion, from the story's point of view, is the following often-quoted exchange:

> "What is REAL?" asked the Rabbit one day, when they were lying side by side near the nursery fender, before Nana came to tidy the room. "Does it mean having things that buzz inside you and a stick out handle?"
>
> "Real isn't how you are made," said the Skin Horse. "It's a thing that happens to you. When a child loves you for a long, long time, not just to play with, but REALLY loves you, then you become Real."
>
> "Does it hurt?" asked the Rabbit.
>
> "Sometimes," said the Skin Horse, for he was always truthful. "When you are Real you don't mind being hurt."
>
> "Does it happen all at once, like being wound up," he asked, "or bit by bit?"
>
> "It doesn't happen all at once," said the Skin Horse.
>
> "You become. It takes a long time. That's why it doesn't often happen to people who break easily, or have sharp edges, or who have to be carefully kept. Generally by the time you are Real, most of your hair has been loved off, and your eyes drop out and you get loose in the joints and very shabby. But these things don't matter at all, because once you are Real you can't be ugly, except to people who don't understand."[9]

Not long after this exchange, the Boy takes the Velveteen Rabbit to bed with him after being unable to locate his regular toy companion. From then on, the story follows the growing relationship and mutual attachment of the Boy and his Rabbit through the various seasons of their friendship. Throughout this narrative, the story is told from the Rabbit's point of view. On three further occasions, the question "What is REAL?" emerges in some form.

On the first of these occasions, the Boy is unable to go to bed without the Rabbit (which is accidentally left on the lawn outside). On returning the Rabbit to the Boy, Nana expresses her incredulity at "all that fuss" over a toy, to which the Boy retorts, "You mustn't say that. He isn't a toy. He's REAL!"[10]

On the second occasion, the Rabbit is propped up in the garden somewhere and is visited by two wild rabbits. They are at first curious, and quietly and cautiously they move closer to the Velveteen Rabbit to inspect him. Then follows a series of exchanges between the wild rabbits and the Velveteen Rabbit. The conversation ends when one of the wild rabbits comment, "He doesn't smell right! . . . He isn't a rabbit at all! He isn't real!"[11]

In the final part of the story, the Boy is stricken with scarlet fever for many weeks and is bedridden, with the Rabbit by his side. When he is finally well enough to leave his bed, the doctor and Nana conspire to dispose of the Rabbit because of their belief that this toy is now "a mass of scarlet fever germs!"[12] The Velveteen Rabbit is replaced with a new rabbit toy, but the story of the Velveteen Rabbit does not end here. The Velveteen Rabbit is stuffed into a sack by the garden shed in preparation for being burned. That night, alone and in grief, the Rabbit reviews all the good times he had with the Boy. At this point, the final question concerning the nature of reality is asked.

Of what use was it to be loved and lose one's beauty and become Real if it all ended like this? And a tear, a real tear, trickled down his shabby velvet nose and fell to the ground.[13]

From the spot where the tear falls grows a flower, and inside that flower appears a being of gold and pearl perfection, the nursery magic fairy. This fairy takes care of all the playthings that have been loved but are no longer needed by children. The Velveteen Rabbit is transformed into a wild rabbit and is taken to Rabbitland to join all the other wild rabbits "to live . . . for ever and ever"[14]

This ending has been the focus of severe criticism, mainly from Faith McNulty[15] and Gerald Weales.[16] McNulty describes the message conveyed by this ending as "sad," "sleazy," and "false as a three dollar bill." McNulty criticizes the story for not providing ideas about positive ways of meeting the "tragedy of lost love and betrayal [without] letting the heart break."[17]

Continuing in this vein, Weales wonders why the Rabbit's magical reward did not extend to his old friend the Skin Horse, a point also noted by Daniels and described as an "inconsistency."[18] Weales also regards the ending as a transformation brought about by self-pity rather than the power of Love. Daniels, pondering these complaints, offers psychoanalysis, "in particular Melanie Klein's theories," as a way of "confronting the question at the heart of the story."[19] The question that Daniels believes to be at the heart of the story is, ironically, the last one described at the end of the story: "Of what use was it to be loved . . . and become Real if it all ended like this?" Focusing on this as the central question of *The Velveteen Rabbit*, Daniels argues that the story is about the "ambivalence toward separation," which is resolved only through "idealization." Idealization (in the form of the fairy) is viewed as a defense mechanism, a way of resolving "the persecutory anxiety involved in feeling oneself abandoned by the 'all-important person'."[20] The story of *The Velveteen Rabbit*, Daniels argues, is an analogue for the inner lives of infants or children. Furthermore, he argues that the Kleinian interpretation is a "standard against which the question it poses can be best understood."[21]

There are several problems with these views. To begin with Daniels's final comment, there are no definitive standards in the semiotic arts save those agreed to by consenting, like-minded readers. As Ann Game argues, textual analysis should not be understood as simply "representation but as itself a writing or discursive practice."[22] In other words, interpretations can be seen as another way of reading or telling the story, and these are subject to the particular cultural characteristics of audience and readership.

The Kleinian interpretation is rather esoteric as a reading and therefore may not resonate with audiences outside psychoanalytic circles. This does not, of course, invalidate the interpretation, but it does suggest that the question of which frameworks are most relevant in understanding the central question of the story may still be open. There are two further reasons for suggesting this.

First, the psychoanalytic reading does not identify or explore simpler social and moral themes, even if only to connect these to the defensive ones of a Kleinian interpretation. And yet, these simpler social themes may be more accessible to the emotional sensibilities of young readers. In this way, these themes may be more influential in shaping the thoughts and feelings of young readers.

Second, the question that Daniels believes to be at the heart of the story is, I think, not the central one at all. On the contrary, the question asked by the Rabbit during that dark night in a sack should be seen as supporting the story's overall theme. If it is seen in that context, the story is also not about the "ambivalence of separation." These points tend to erode the credibility and therefore the persuasive power of a psychoanalytic interpretation of death and renewal in *The Velveteen Rabbit*. In support of these observations, I will first discuss what I believe to be the central question of the story. I will then show how the Rabbit's death scene continues and extends the main theme. Finally, in the context of that discussion, I will address the other concerns of McNulty and Weales.

A Sociological Reading

THE CENTRAL QUESTION OF THE STORY

The central question of *The Velveteen Rabbit*, the focus of both the plot and the recurring symbols, is "What is REAL?" This question constitutes the only prolonged verbal exchange in the book, between the Velveteen Rabbit and the old Skin Horse. This question is developed further in exchanges between Nana and the Boy, between the wild rabbits and the Velveteen Rabbit, and, finally, of course, in the scene toward the end between the Rabbit and the magic fairy.

In the first major treatment of this question, the Skin Horse is careful to point out to the Rabbit that being Real has nothing to do with having "proper insides." It is not, therefore, an issue of biological or mechanical authenticity. It is not about composition, makeup, or even pedigree in any absolute and reductionist sense. Rather, Real is "what happens to you," in other words, how you are changed by *interaction with others*. It is the perceived enjoyment and exchange of affections, the wear and tear (literally) of social interaction, that creates the process of change. Real is a social and physical process of change that can transform a toy into a friend, from simple object to prized play relation. And as their relationship develops, the child grows and the wear and tear on the toy becomes the measure of the toy's social value in that mysterious process.

This idea of reality as the social construction of identity through friendship is further developed in the exchange between Nana and the Boy. When the Boy asks Nana to refrain from referring to the Rabbit as simply a toy and vehemently asserts its realness, he is reaffirming this social relationship between two *social* identities, himself and the Rabbit. Real means "loved" or "precious." It means "real to me and for my purposes." The Boy's relationship with the Rabbit fulfills all the vital functions of friendship and is therefore, for all intents and purposes, real, that is, socially real.

The Boy's reality, then, is a social reality defined by attachment. This is a bond cemented and maintained by the shared experiences of companionship, a companionship believed to be reciprocal. Compared to this reality, the material reality of the world is not reality at all, at least not for the boy.

The gardens of play are stages, and the wild rabbits are part of those stages as props and scene pieces, but they do not constitute the real action. Real action follows the Boy's ideas in a social world that the Boy creates. Part of the exchange with the wild rabbits about what is Real is revealing here:

> "Why don't you get up and play with us?" one of them asked.
>
> "I don't feel like it," said the Rabbit, for he didn't want to explain that he had no clockwork.
>
> "Ho!" said the furry rabbit. "It's as easy as anything." And he gave a big hop sideways and stood on his hind legs.
>
> "I don't believe you can!" he said.
>
> "I can!" said the little rabbit, "I can jump higher than anything!" He meant when the Boy threw him, but of course he didn't want to say so.[23]

These wild rabbits appear to be a major contrast to the social reality of the Velveteen Rabbit until one realizes that, their physical appearance notwithstanding, these rabbits talk. The poetic license taken in this respect further highlights that the world of the Boy and his Rabbit is a socially constructed world in every major respect. This is in no way an inferior world because, like our own empirical world, the material is necessary but not the authoritative basis with which we build our meanings. Furthermore, the subtle emphasis on the social nature of things becomes an increasingly important perspective that is developed to complement and support the story's unfolding theme.

The final exchange concerning the nature of reality is between the fairy and the Rabbit. The Velveteen Rabbit was Real to the boy but now would be made "Real to everyone." But did the fairy mean Real like the biologically real wild rabbits or Real as in the Boy's and Skin Horse's meaning? Initially, the evidence appears conflicting.

On the one hand, Rabbitland seems to be the place where all the wild rabbits live and play. The final scene describes the Velveteen Rabbit, now a wild rabbit, visiting the Boy from a distance. They exchange looks and the Boy, just for a moment, is reminded of his old Velveteen Rabbit from the markings on the real rabbit he sees.

On the other hand, the Velveteen Rabbit is meant "to live . . . for ever and ever" with them. If they are biologically real wild rabbits, "for ever and ever" amounts to about eighteen months.[24] This hardly amounts to immortality. Similar to the little figure of the Velveteen Rabbit, then, this suggests that Rabbitland has similarities to an actual rabbit, or rabbit culture, but in fact departing from it in important ways. Similar to the Spiritualist's Summerland, then, here is a transformation of some aspects of shape and function but preservation, or at least remnant persistence, of others.

Once again, then, the physical shape of things plays only a minor role in the social construction of their meaning and value for all the participants. Far more important are the processes of identity construction and transformation derived from Love rather than the substance of any single event (e.g., hurt, death, discomfort) or inherited characteristic (e.g., polished eyes, shiny fur). The fairy's attitude to Real, then, seems consistent with those asserted by the Boy and the Skin Horse earlier in the story, and also in the spirit of the images of biologically real rabbits that talk. The Velveteen Rabbit then becomes "real to everyone" only insofar as the wild rabbits themselves are understood to be real in William's sociable, chatty, but anthropomorphic portrayal. More than this, though, the Velveteen Rabbit will also become socially real to the readers and listeners of the story. When boys and girls encounter biologically real rabbits in the future, each may ask: Whose rabbit do I see before me now? Which boy or girl has made this rabbit real? The relationship between animals and children is no longer viewed merely in biological terms but, more importantly, in social ones. And so it is in the pursuit of the question "What is Real?" in each of these major scenes that the main narrative energies are absorbed and dedicated.

The death scene is the final dramatic and supporting contribution to this question of "What is Real?" The scene amounts to a tale of personal survival and rebirth for the little rabbit. Death in this story is not presented as the final triumph of loss and despair over attachment and love but rather as the final evidence that loss and despair are unable to surmount what was not theirs to surmount in the first place. In this respect, the narrative reality of *The Velveteen Rabbit* is not the triumph of a materialist theory of existence exemplifying the inevitable competitive edge that death has over all life.[25] And perhaps it is this departure from conventional psychoanalytic wisdom that attracts so much suspicion and disappointment from those so inclined.

Instead, Margery Williams's message is for children *not* to be seduced by the confusing and irregular appearance of love and loss in life. They are encouraged to realize that love and loss are constant riding companions in the rough and tumble of life's unpredictable splendor. Loss is not triumphant. The more sophisticated message is that love is something that can emerge from, and triumph over, the complex and unpredictable nature of life's changing fortunes.

For example, at the beginning of the story, we witness the attachment between the Boy and the Rabbit on Christmas morning. Loss follows some hours afterward as the boy's interests shift to other toys. This absence lasts a long time and leads to the friendship between the Rabbit and the old Skin Horse. This friendship, in turn, ends when the next period of attachment to the Boy begins. The initiative gained by this attachment is the centerpiece of the story and is overtaken by loss only after the scarlet fever. In this context, the death or separation scene toward the end of the story should be read not as idealization or denial but as a narrative device. But this device does not support some simple biological conception of life that, as I have argued, is never shared by the main players. Rather, a concept is further developed of an existence in which change and transformation are paramount and integral to the story and, indeed, life itself.

The story in general, then, impels its young readers and listeners to treat all their toys as socially real and to care for them. The story's images suggest that toys are dependent and vulnerable creatures (like children themselves). The toys eventually change, becoming almost unrecognizable from their original selves (again, like growing children). The exhortation is to care for each other, to show kindness and affection, until the time when everyone becomes Real. Children become socially real when they grow up to be adults, and toy rabbits become real when they lose their shape and are no longer required as playthings.

The complex idea which is essential to convey to young readers at this point is the inevitable involvement of everyone in this process. Love, and the care and kindness that are its hallmarks, do not guard against loss and death. Love, paradoxically, creates these experiences. Without attachment it is loss that cannot exist, and without love it is death that has no dominion. This is why Real does not happen very often "to people who break easily, or who have sharp edges, or who have to be carefully kept."

The Continuing Theme

To convey this theme one final and impressive time, the Velveteen Rabbit undergoes a death that has remarkable similarities to our understanding of the NDE. Death begins with the final separation from the Boy. The Rabbit is stuffed into a dark sack, a symbol coincidentally used by Tolstoy to convey the onset of death.[26]

> And so the little Rabbit was put into a sack with the old picture books and a lot of rubbish, and carried out to the end of the garden behind the fowl house. That was a fine place to make a bonfire, only the gardener was too busy just then to attend to it. He had the potatoes to dig and the green peas to gather, but next morning he promised to come quite early and burn the whole lot.[27]

Reemerging from the sack, the Velveteen Rabbit reviews his life and all his major social events and relationships.

> He thought of those long sunlit hours in the garden—how happy they were— and a great sadness came over him. He seemed to see them all pass before him, each more beautiful than the other, the fairy huts in the flower bed, the quiet evenings in the wood when he lay in the bracken and the little ants ran over his paws; the wonderful day when he first knew that he was real. He thought of the Skin Horse, so wise and gentle, and all that he had told him.[28]

He then encounters a special being in the form of a fairy and is taken to another place and another life.

> And presently the blossom opened, and out of it stepped a fairy.
> She was quite the loveliest fairy in the whole world. Her dress was of pearl and dewdrops, and there were flowers round her neck and in her hair, and her face was like the most perfect flower of all.[29]

Here the world is both different from and similar to his former world but is a place where immortality reigns. Daniels[30] and Weales[31] before him imply that this is an ending not shared by the Skin Horse. However, there is no evidence in the story that the Skin Horse was at the end of its use. After all, the Skin Horse lived in the nursery with other toys, perhaps forgotten but not discarded. The Velveteen Rabbit, on the other hand, faced death by incineration the next morning. This eventuality was carefully created and described for the Rabbit's unique role in the story as the Boy's favorite toy.

The death narrative in the final part of the story continues the theme of the triumph of love in the face of unpredictable events. In that context, Margery Williams's death scene supports her philosophy of life. (This is not a surprise given our knowledge of her family's tendency to move so frequently during her childhood.)[32] But Williams did not make the death scene thematically central. As a plot climax, it has always played a supportive role to the social and poetic themes developed throughout the rest of the story. How does this death scene, highlighting the problem of loss, actually privilege the theme of love as developed by the story so far?

First, the Velveteen Rabbit's feeling of loss is a consequence of his love for the Boy, a love that causes him to pine and despair at the prospect of permanent separation. The appearance of the fairy, in turn, is a consequence of the Boy's love since, according to the fairy, she only takes "care of all the playthings that [children] have loved."[33] Love, therefore, and particularly the fate of love when confronted by loss, is the concern linking the fairy and the rabbit.

Second, both the fairy and Rabbitland as images of death and renewal deliver an important message that also celebrates the general theme of the book: that love can triumph over loss. Love does not diminish with change but can continue in transformed ways in spite of it. This is because both children and their toys live on in their futures as memories and dreams, the past a living social presence and an influence on the present.

In this context, Rabbitland demonstrates the importance of the *social* rather than material nature of love. It continues the theme that love's heart does not miss a beat even if "most of your hair has been loved off, and your eyes drop out and you get loose in the joints and very shabby."[34] In other words, neither age nor death, as material appearances that can deceive, should be permitted to define the limits of emotional and social attachment.

Finally, the portrayal of Rabbitland as an empirically real place has parallels with retelling NDE accounts. Carol Zaleski argues that such portrayals help listeners and readers connect future possibilities with present circumstances and actions in one's imagination.[35] This facilitates the shaping of present priorities and attitudes, and this is exactly the desire of many storytellers, including Margery Williams. Rabbitland, then, is a narrative device that helps young readers understand how love may cope with loss.

The message that love is a social relationship not necessarily determined by material conditions is first expressed by the Skin Horse in the nursery. And it is this message, in this final form, that makes the death scene a convincing narrative about the power of love and not, as Daniels argues, about separation.

This message encourages the development of a more sophisticated idea about relationships that can facilitate the ability to transcend loss. This is because, as Williams reminds us, love enables our attachments to remain with us as living influences on our thoughts, feelings, and actions. This is not a denial of death but rather a moving beyond it, a refusal to be defined and immobilized by it. Objects and experiences are personalized through the processes of attachment, and this, in turn, creates the intellectual and emotional content of social life. The importance of this principle is echoed by Margery Williams herself when she remarks that "it is through imagination that a child makes his [sic] most significant contacts with the world around him, that he learns tolerance, pity, understanding and the Love of all created beings."[36]

It is true that the idea of loss has a presence, but it is neither the final nor the dominating theme of the story. This is because, as I have argued, its narrative context is the transforming power of love and its survival, and even triumph, in the face of personal and social change. McNulty is therefore incorrect when she complains that "the book gives no hint that there is any way to meet the tragedy of lost love and betrayal." To put the matter rather simply, no love was lost in *The Velveteen Rabbit*. Furthermore, there was no betrayal, certainly not from that "all-important person" the Boy.

Although the little rabbit was discarded by the Boy's guardians and not, it should be noted, by the Boy himself, the love between them continues in both of their subsequent lives and memories. Therefore, the story's theme is not about the "ambivalence of separation" but rather about the robust ability of love to transcend separation even in death.

Nor is the final transformation of the Rabbit brought about by "self-pity," as Weales asserts. Rather, the author shows, and shows consistently, that the love between the Boy and his Rabbit does not simply die with the material absence of one of them. This is not a case of out of sight, out of mind. Rabbitland is a "life after death" for the Rabbit, and this encourages young readers to entertain the social presence and influence of a relationship long after its material existence has come to an end.

Acknowledging the ongoing influence of past relationships on present ones is not easy for most adults, let alone children. The near-death imagery of the death scene facilitates empathy for those social but invisible influences in all our lives. Far from being motivated by self-pity, the images of death and renewal are drawn for a specific pedagogic purpose: to provide effective and imaginative support for the story's final message—real love continues even when the beloved is taken away.

And so, to that end, Margery Williams writes, using the seasonal language of the grieving heart:

Autumn passed and Winter, and in the Spring, when the days grew warm and sunny, the boy went out to play in the wood behind the house. And while he was playing, two rabbits crept out from the bracken and peeped at him. One of them was brown all over, but the other had strange markings under his fur, as though long ago he had been spotted, and the spots still showed through.

And about his little soft nose and his round black eyes there was something familiar, so that the Boy thought to himself:

"Why, he looks just like my old Bunny that was lost when I had scarlet fever!"

But he never knew that it really was his own Bunny come back to look at the child who had first helped him to be Real.[37]

Implications for Interpretation of the NDE

Examining images of death and near-death in children's literature has several implications that differ somewhat from those of adult fiction.[38]

First, many social and literary commentators on children's literature, as we have seen, are concerned when images of death turn into tales of re-birth. Drawing on materialist ideas about death and having tolerance for little else, these reviewers look to a psychoanalytic theory for an explanation. But that reaction and writing itself, through its own failure to grasp a plurality of ideas about death, overlooks the opportunity to explore different cultural meanings. This ethnocentric psychoanalytic response problematizes any concept of death associated with ideas of survival and rebirth. These different notions of death are considered deviant, both socially and psychologically.

That process, I believe, undercuts the importance of understanding what death means to different people, in different times and places, with different social experiences of it. A sociological reading of *The Velveteen Rabbit* reveals the value of providing alternative models to psychoanalytic ones that portray death as annihilating and final. We can show that unlike Freud's medical view of death as a winner who takes all, symbolic representations or social experiences of death are always context dependent. This is because sociological readings assume that a meaningful understanding of death derives from its logical relationship to other meanings that emerge from a society or story. Images of death in *The Velveteen Rabbit* should be understood in terms of their narrative context. Concepts of death in psychoanalytic discourse should be seen in their historical and political context. And as Ring has recently argued,[39] personal NDEs should be read in their biographical context in much the same way that Zaleski views the public discourse about NDE in the context of religious history.[40] In these ways, no meaning of death need foreclose on another, whatever scientific rhetoric is used to justify and rationalize such a monopoly.

Because tales of rebirth in the face of death are commonly subject to the analytical dominance of materialist and psychoanalytic paradigms, at least in the social sciences, it is important to provide alternative readings. This is because, as Jean-François Lyotard argues, the social and intellectual value of such rewriting and retelling helps "invent allusions to the conceivable which cannot be presented."[41] And that exercise, in turn, opens up intellectual spaces where we can entertain and examine, with a range of methodological tools, the plurality of ideas about death that is part of our diverse human inheritance.

Second, Melvin Morse has recently demonstrated how childhood NDEs are strikingly similar to adult versions.[42] He suggests that since children are not fully socialized creatures, cultural conditioning may play little or no part in the NDE. Aside from the methodological problems of retrospective recall in many of these cases, the conclusion is perhaps premature for another reason. Children do not become socialized suddenly, despite the desperate wish of certain parents. Socialization is incremental and gradual, dependent as it is on processes of physical and social development. The similarity of childhood NDEs to adult versions does not suggest that cultural conditioning is unimportant. Rather, it can just as readily be interpreted as evidence that such conditioning occurs rather early.

The Velveteen Rabbit contains many of the classic images of the NDE. However, it is not unique in this respect. In *The Wizard of Oz*, for example, Dorothy is transported inside the "tunnel" of a tornado to another place, where she meets the "good witch of the east."[43] In *Alice in Wonderland*, Alice begins her adventure by a long fall down a dark rabbit hole.[44] Children's literature is replete with tunnels, extraordinary beings, life reviews, flying experiences, and tales of reunion.[45]

At least in the literate circles of industrialized nations, children have a wealth of imagery to draw on to help them make sense of everyday and extraordinary social experiences. These stories, like *The Velveteen Rabbit*, have a long history, and they are read by or to children at very early ages. I do not argue in any reductionist way that childhood NDEs are merely a product of these fantasies. But I do draw attention to the fact that children's stories provide one source of cultural imagery (and there may be others) that may assist children in making sense of their NDEs.

Without some cultural materials it is doubtful that anyone, child or adult, would be able to communicate even minimally, let alone make personal sense of, foreign experiences such as the NDE. Individuals of any age are not culturally neutral. Therefore, the descriptive narratives of NDEs should be understood in terms of the language of that communication, a point discussed at some length by Zaleski.[46] In that same work, Zaleski argues that

"When one judges a symbol, one cannot say whether it is true or false, but only whether it is vital or weak," and this depends on its "capacity to evoke a sense of relationship."[47] Few of the millions of readers who have read or heard *The Velveteen Rabbit* since its appearance in 1922 accept the literal meaning of Rabbitland. But even fewer would argue against the social message of the story. Those of us who do not believe that love can triumph over loss wish nevertheless that it could. And that very desire to look again at the universal mystery of love and death continues to breathe new life into the story of *The Velveteen Rabbit* and to ensure that our relationship to tales of rebirth, of transcendence of death, remains vital and relevant.

9

CRISIS AND MEANING

From our study of *The Velveteen Rabbit*, it should now be apparent that the key to understanding the methodology underlying all the previous chapters is to recognize that they involve some rebuilding of context. Previous images of the NDE taken and discussed out of context are returned, placed back into the social and cultural fabric from which they were removed, discussed, and dramatized by medicine and psychology. Let me review these insights briefly.

When clinical NDEs are compared the world over, they do not follow the Western form in all important respects. Western NDEs are not universal. When the clinical NDE is examined more closely, its psychological elements viewed against the background of their social origins, NDEs can be seen as unusual experiences in unusually stressful circumstances. NDEs are not simply medical phenomena. Away from the closeups of individual case examples encouraged by early anecdotes, we are able to see more broadly that, far from being unpopular, the NDE enjoys enormous popularity, interest, and support.

And the academic explanations—obscure and impatient, demanding in spite of these uninviting and authoritarian qualities—turn out to be, with slow and patient reading, a disappointment. Placing them in the context of their declining authority, and sorting out polemical psychobabble from substance, we are not, unfortunately, left with the pure technical offerings that these sciences promise. Rather often a partisan, ethnocentric, and conservative narrative is revealed, one that resists an appreciative, inductive gaze for things new and unusual—an ironically unscientific approach if there ever was one.

In this final chapter, I will draw these threads of analysis together and, for one last time, attempt to provide a larger context for their understanding.

Both NDEs and the community and academic reactions to them can be understood as a crisis in meaning. For NDErs, the experience is one of a number of personal crises that most, if not all of us, may endure during our lives. For groups such as scientists, coming to terms with the meaning of the NDE may mean adjusting to the idea that they have failed the general community that has supported them, particularly in matters important to that community. Technical definitions of death have little meaning and even less relevance to ordinary people. In this way, people have been locked out by a tradition of instrumental reason designed for technological innovation rather than ethical and social anxieties. This failure has contributed in no small way to the decline of the social and philosophical relevance of science, now widely seen as merely sites for medical cures and gadget development.

To illustrate these points, I will first discuss what is meant by the term *crisis* and then examine its meaning in terms of personal experience. In doing so, I will redraw Moody's classic vignette of an NDE, integrating our new sociological understanding of the experience into its images. In the final section of the chapter, I will discuss the possible reasons why the academic reaction is so at variance with the community one.

Personal Crisis

Human life is an assortment of experiences that are constantly changing. And yet, despite the banality of this observation, people have great trouble coping with these changes. The reasons have to do with the habits and rituals we develop to chart our way through the social mess we call life. Then, as we relax into these habits and rituals and take their meanings for granted, we mistake them for life itself. As long as the challenges we encounter daily are minor, this deception works as a great time saver, if for no other reason. One traffic jam or family squabble is much like any other, and recognizing the pattern helps us to deal with them efficiently. Habits and rituals, then, are adequate first responses to the familiar. In the confined social spaces of family, school, and work, habits and rituals proliferate, their personal expression adding to our understanding of each other's individual differences and characters. In this way, the original banal observation of the changing nature of life is obscured, indeed eclipsed, by the complementary observation that life is heavily regulated and made predictable by the organization we impose on our daily lives.

When we are confronted by the unfamiliar, we are challenged. In this situation we must solve the problem or back away from it if we can. Arguments with friends, increasing workloads, or completion of the writing of a poem are minor, everyday examples of this challenge. Each fragment of the familiar has a kernel of the unfamiliar that gives all situations a slight tension. But for nearly all of us at some stage of our lives, this fiction of the predictable sooner or later unravels because life is invariably greater and more unpredictable than our best-designed plans and responses. A divorce, an unaccountable depression, job loss, the death of a friend, a life-threatening accident or illness, a foreign environment—all these and many other events can precipitate a crisis.

A crisis has been described as a transitional period in life, a turning point in values or attitudes when confronted by an "unfamiliar obstacle in life's path."[1] The personal resources one has to cope with this obstacle may be found wanting, and new ways to cope need to be found if one is to meet the challenge. In the period during and after the crisis, one's whole way of understanding life may need revision because a major part has been found wanting. The meaning and value of one's life may be called into question. Much of the impact of a crisis comes from previous habits of thought and from the assumption that one's understanding of the world is adequate for crisis management.

The formidable power of crises is also derived from the fact that our communities are fragmented systems that do not allow us to learn about unpredictable things easily. Reading, TV viewing, and gossip must suffice until experience can take their place. Furthermore, not everyone survives experience. These cultural circumstances encourage the view that crises happen to others. Crises are unusual happenings which make the TV news instead of the steady stream of disruptive and rupturing change which underlies the body, human relationships, economies, and travel.

The concept of crisis then, described in this way, must be assumed to be neither an abnormal psychological or social situation. This is because it *does* occur regularly, despite being conceived and rationalized as uncommon by our daily habits of thought and practice. Crises are a usual part of life, however unwelcome that thought may be. In this way, NDEs are important to examine not simply because of their death imagery, but also because they are yet another crisis that disrupts our lives despite our best efforts to avoid them.

Major personal crises have several features that make NDEs easily understood as part of the human experience. In Chapter 3, I described how the clinical NDE is really a subset of social NDEs. Here, however, I want to take this idea further and suggest that social NDEs themselves are part of the normal pattern of human crises. With a little revision, the social features of the NDE described in our study of castaways can also be seen to underlie the general experience of crisis. The psychiatrist Beverley Raphael argues that both crisis and disasters are usually events or circumstances that overwhelm or test the capacity of communities and individuals to cope. They may be of gradual or rapid onset. Crises are of interest both psychologically and sociologically because they represent *stressors* for the individual and society. However, Raphael assumes a hemostatic model of crisis and disaster, arguing that

> there is also, by implication, a process of recovery—a return to some previous or new equilibrium.[2]

This, of course, cannot be true. Many a disaster or personal crisis has seen the permanent destruction of innumerable communities and individuals. As I observed earlier, some people do not survive experience. For them, there is no recovery or "new equilibrium." Nevertheless, the idea of crisis suggests that we look to the physical or social world for the stressors that may be prompts to further unusual psychological and social experiences. People in crisis may experience their lives unfolding in the following way:

Unusual physical or social experiences can provide a deep, disturbing, or disorienting sense of

Social separation. This can be viewed as the obstacle to overcome, a problem associated with a

Perceived danger to the self or the self-concept and hence

A sense of helplessness and uncertainty of survival. This might prompt

A tendency to review of one's life and

Unusual perceptual experiences. This can lead to

major personal reorganization or breakdown.

These features of crisis have been shown to fit our earlier discussions of clinical and social NDEs. However, it is now time to show that these features also apply to many other crises that do not threaten one's physical life. I will demonstrate once again, not with NDE cases but with another crisis experience, how remarkable social, psychological, and paranormal experiences can persist, well beyond the confines of medicine and the process of dying. Only when we continually see this broad range of experiences from other events in life can we begin to understand how the NDE has been taken out of a much broader context of human experience.

The Experience of Bereavement

Bereavement is a good example of crisis. Bereavement, more common than being shipwrecked or undergoing an NDE, is a crisis that few of us will escape. What follows is a summary description of bereavement that most health and welfare professionals will easily recognize.

UNUSUAL PHYSICAL AND SOCIAL EXPERIENCES

Experiences not encountered under ordinary circumstances may dominate one's life. Widowhood, and the deep sense of loss and loneliness that accompany this event, can be one of these. This critical incident can bring further unusual physical and emotional experiences: prolonged and uncontrollable weeping; depression and anxiety; anorexia; insomnia and gastrointestinal disorders such as prolonged diarrhea or constipation; lethargy; and so on. The social experiences are similar to those of persons who have undergone divorce or separation: loneliness and social isolation; stigma; financial difficulties; increased child care or work responsibilities; the pressure to learn or relearn social skills for handling members of the opposite sex; and so on.

SEPARATION

The experience of separation created by another's death can begin early, starting with a chronic illness that restricts joint activities. Of course, separation can be sudden, as in an acute illness or accident. But the strongest sensations of separation often occur after the death of a loved one. Bereavement is a long process, not very well understood until recently; it may extend over many years.

SENSE OF HELPLESSNESS AND UNCERTAINTY OF SURVIVAL

The social and emotional burdens are so often great that they create moments of despair and desolation, the feeling that one cannot continue to live without the person one has lost. One is unable, at times, to see a way forward alone. The death of someone so loved and so important can threaten one's very existence, robbing one of a reason to live. How can one go on without that person?

UNUSUAL PERCEPTUAL EXPERIENCES

Then, one afternoon while cleaning the kitchen mindlessly, the person looks at the breakfast table, only to see the deceased sitting there and smiling. They say they are fine, and there is no need to worry for them. And then, as suddenly as they appeared, they disappear. At other times one seems to hear the voice of the deceased offering comfort or advice. Sometimes one addresses the voice, and a conversation ensues.

LIFE REVIEW

A constant tendency to reminisce, to remember the shared life, is an integral part of the bereavement experience. This fuels the power of grief and is also fueled by it in a seemingly endless cycle of sorrow and pathos. As the psychiatrist Butler observes, "one tends to consider the past most when prompted by current problems and crises."[3] Eventually the bereaved person may not be able to perform his or her usual tasks and may need assistance in coming to terms with life without the deceased.

MAJOR PERSONAL REORGANIZATION

More often, life does resume steadily, either through one's own effort or because the demands of other loved ones prevent one from being entirely swallowed by grief. Major reorganization of life has begun and continues.

Others know, even empathize with, the bereavement experience. The bereaved person tries to tell of conversations with the deceased but attracts sideway glances and sympathetic but not understanding comments. Sometimes he or she is able to meet others who have undergone this experience, and sharing becomes possible. Bereaved persons choose their confidants carefully. Still, life is never the same, and the experience profoundly affects their lives always, particularly their understanding of death and life's transitions.

We are not used to viewing the NDE in these terms, although the NDE fits this description well. One reason for this oversight must be that we associate the constellation of NDE characteristics only with experiences near death. Most important, we have been persuaded to see the NDE as a dying experience rather than as the life experience that it actually is. However, as we have seen in the example of the castaways, one can feel being in danger, remain physically well, and experience many of the features of the clinical NDE. But this one example may be too remote for some people, preventing them from making the connection between the NDE and other more common crises. In this context, bereavement is not only a widespread experience, it also provides an excellent demonstration that the rather remarkable features of the NDE are not confined to it.

One insight to emerge from the bereavement narrative is that the interpretation of an experience cannot be separated from the way it is recounted as a story. Because many texts on bereavement have a therapeutic purpose, their focus is largely emotional. Sadness, anger, experiences of shock or denial, and depression are privileged in these kinds of narratives. Medical narratives of the bereavement experience may also emphasize the bodily changes and illnesses of the bereaved. Social work narratives may emphasize aspects of social adjustment. Parapsychological narratives are more likely to privilege aspects of bereavement such as visionary encounters with the deceased.

All such narratives are no less true for their respectively differences, but it is important to note that our understanding of such experiences, whether bereavement or the NDE, are very much shaped by such narratives, especially if we have little experience of our own.

Similar narratives can be constructed for other crisis experiences. For example, falling mountaineers often experience life review, as do trapped miners; shamanic initiates often experience OBE, as do victims of child abuse and torture; Antarctic explorers and those lost in a variety of wilderness situations often meet deceased relatives or supernatural beings; those who undergo a sudden religious conversion also encounter a light, experience a feeling of peace, and sometimes even witness supernatural vistas. Most of these people are never the same afterward, no longer viewing life as stable, predictable, or pedestrian.[4] Many lives are dramatically transformed in both positive and negative ways.

All these experiences point up the great diversity of psychosocial elements in human crises. For example, some people experience only an out-of-body sensation, others only a life review. Still others encounter many features of what we have come to know as the "full" Western NDE experience: tunnel sensation, OBE, life review, meeting deceased or supernatural beings, and altered social values and attitudes. The bereavement narrative above has been constructed this way, drawing on the known but broadly variable features of the bereavement experience.

However, it must be remembered that the full image of the NDE is an artifact of the composite picture put together by Moody and repeated endlessly in the NDE literature. Few people actually experience all of these images. This situation, remember, also characterizes the castaway experience and the clinical NDE. In Ring's study of clinical NDEs, for example, 60 percent of NDErs experienced a sensation of peace but only 23 percent "entered a darkness" and only 10 percent experienced supernatural scenes.[5] People who are bereaved are probably more likely than NDErs to see deceased relatives or friends. (It is estimated that up to 66 percent of widows and 75 percent of parents who lose a child experience contact of some sort with the deceased.)[6] As Moody himself cautions:

> I have found no one person who reports every single component of the composite experience. . . . [7]

Our understanding of the NDE has been shaped not by the diversity of NDEs but by a collective portrayal of features seldom occurring all together but immortalized by Moody's medical vignette. There are three serious and interrelated consequences of this situation.

First, the narrative model defines the limits of the experience through its medical characterization, helping to overidentify its social features with the experience of illness, injury, and resuscitation. Also, by linking five or six features that often do not occur together, it reduces the breadth and complexity of NDEs while privileging a single seldomly occurring pattern of the NDE.

Second, while the report of tunnel sensation may be of minor interest alone, or the OBE sensation may be old news, the arrangement of these elements together makes people sit up and take notice. Engaging though this device is, it is more a theoretical step than one driven by the empirical evidence. This is because underlying the composite model is an attempt to theorize that the elements of tunnel sensation or OBE sensation may be part of a particular scenario. They may constitute stages of dying. However, this conclusion is derived prematurely at the expense of a wider, more careful examination of human experience.

Third, the fame of Moody's narrative has allowed his interpretation or model to dominate our image of death, privileging not simply one pattern of dying experience but also one form of this crisis. This facilitates the view that the NDE contains most of these unique social and psychological elements. Since this is plainly not true, such a narrative, though broadening our understanding of death, ironically restricts our understanding of human crisis. Such images confine our understanding of the remarkable to the misty edges of death at a time when contemporary studies of crisis show that such events are a regular feature of most people's conscious lives.

Overall then, the problem with Moody's composite picture is that anyone who demonstrates some of its features is deemed to have undergone an "NDE," whether near death or not. In this methodological way, experiences such as imprisonment and multiple personality disorder, if they include OBEs or other mystical events, have been linked to NDEs instead of to the experiences of life crisis that they really are.

The important features of the NDE that have captured the popular imagination are commonly associated with crises such as shipwreck or bereavement. And because crises are normal events, their social and psychological elements should be seen in this context. With these points in mind, let us now redraw the clinical NDE vignette as one of general human crisis.

The Experience of Crisis

Persons are separated from others, either physically, socially, or emotionally (severe depression, shipwreck, life-threatening illness or injury, bereavement, imprisonment, etc.). At some point during this experience, they begin to realize that they may not be reunited with those they care most about. At the same time, they undergo unusual physical and social experiences. They feel a sense of threat, even danger. They begin to see their situation as others might and begin to assess its consequences and possibilities. The threatening circumstances prompt them to review the meaning of their lives.

Unsure of their chances of survival, or even of whether help will be timely or effective, they begin to reflect on their surroundings. They find that, if nothing else, they are at least alive for the moment, despite certain physical constraints and disabilities. Occasionally, other things begin to happen. They may experience in-

credible phenomena, for example, OBEs or visits from other beings or spirits. Sometimes the visits involve the sensation of a presence. At other times, they are realistic, apparently physical, visits from religious figures or people thought to be dead. Eventually, after the passage of time, either through their own efforts or those of others, the crisis ends. The persons who have lived through it take on new attitudes and values, experiencing a personal reorganization that gives them a new, sometimes positive, outlook on life. They sometimes attempt to tell others, but the paranormal or mystical aspects of the story are sometimes omitted for fear of ridicule or stigma. They choose their confidants carefully. Still, the experience affects their life profoundly, especially their views about death and life crises.

Crises show us all, at some time, the need to review our taken-for-granted meanings about the world—about work, relationships, health, money, certainty and uncertainty. They prompt us, usually with considerable force, to change those meanings so that they can provide greater, perhaps truer, understanding.

The difficult task of altering a major part of our taken-for-granted meanings becomes essential if we are to transcend our troubles rather than being destroyed by them. Crises, in this sense, are the movers and shakers of our most deeply held meanings. And, as the broad features of the preceding vignette indicate, almost every day a war, a natural disaster, the loss of one's job, or the end of a relationship brings the experience of crisis to more and more people.

Although there is little doubt that personal insights, mystical experiences, and unusual perceptual vistas may be seen near death, such revelations are not confined to the experience of dying. The experience of life, pushed and pulled by crises throughout its course, is itself a revelatory experience. Unfortunately, this is easily forgotten when the literature on death is not put in its proper context—life.

Science, Context, and Meaning

Life may be the proper context in which to examine crisis experiences, but most current debate about the NDE continues against a backdrop of controversy over the meaning of death. To recap: adherents of traditional religions have offered philosophies of death that are either poorly developed or peppered with anachronistic imagery. Clinical medicine is unable to define death and unable to know how to act in matters regarding death and terminal care. The biological and psychological sciences seem interested only in an esoteric cellular and molecular view of death so abstract, so divorced from common personal experience, that they are widely ignored. Basically, everyone is unhappy about the topic of death.

In this situation, ordinary people have looked to their own experiences of death—in bereavement, in the many experiences of everyday loss, and in the close encounters of accident and injury. There they have attempted to make connections between circumstances, personal reactions, and meaning. In other words, personal contexts of experience have been important to them. However, they have received little if any help from medicine and psychology—ironically, the very sciences from which one might expect assistance.

Confronted by this seeming indifference, many people have turned to New Age sources to help them review the many arguments about and experiences of death in the broader contexts of religion and philosophy. This turning away from science in general, and medicine in particular, has annoyed the fan clubs of science—the skeptic organizations. Skeptics have preferred to see this critical move away from scientific rationalism as a weakness in human nature rather than as a reaction to cracks in the modern scientific dream.

Meanings of death drawn from human experience have not interested many of those engaged in scientific exploration of the NDE. Neuroscientific explanations, in particular, seem indifferent to the relationship between context and experience, preferring to concentrate on fitting the NDE into preexisting theoretical problems. With a few exceptions already noted, they seem incapable of linking their meanings to a set of workable social psychological ones. And yet, such a link would preserve personal testimony while leaving the philosophical problems to others with more appropriate intellectual equipment.

I have already explored the many complex reasons that draw nonacademic communities to the NDE. In that discussion, I have examined parts of the NDE that attract them and parts of the academic reaction that repel them. Before closing this chapter, I will explore some of the major reasons why the academic reactions to the NDE seem to be fixed on a philosophical course so at variance with the broader cultural reaction. Rather than taking a closer look at the contexts of experience, current academic work appears to be in a steady flight from them.

Under these circumstances, then, we are compelled to ask certain questions of the academic community: Why do they fail to consider circumstances and contexts and, instead, continue to fetishize the psychological responses of people from only one society (Anglo-European) and one circumstance (accident, injury, and surgery)? Indeed, why have the major responses to the NDE from medicine and psychology come from neuroscience, clinical psychiatry, and psychoanalysis? Why haven't more appropriate responses emerged from, for example, social psychology, transcultural psychiatry, medical anthropology, or public health research?

Part of the reason has to do with the successful medicalization of NDE imagery within the academic setting itself. Although this medicalization is counterbalanced by New Age perspectives in lay populations, such perspectives hold less appeal in the academy. A highly secular institution, the academy has been one of the major strongholds and advocates of the traditional scientific paradigm, especially in this century. In particular, there is a widely held belief in the humanities and social sciences that NDEs belong to the province of medicine and/or psychology.

For example, in a recent textbook on the sociology of death and dying, Michael Kearl discusses the NDE entirely as if these were simply medical controversies.[8] More space is devoted to skeptical and medical explanations than to concepts and theories in the author's own social science traditions. The final section of his treatment of the NDE links its imagery to Ron Rosenbaum's view of it as part of pop science cults "which make dying sound like a lovely acid trip (turn on, tune in, drop dead)." Still declining to take the reports seriously, indeed conflating the distinction between reports and beliefs about the NDE, Kearl asserts:

> As matters stand, reports of (and beliefs in) near-death experiences have the same validity status as religious portrayals of the hereafter: both are matters of faith.[9]

In this way, Kearl exhibits the simple territorial understanding of the NDE in much social science literature, such as it is: that which cannot be explained by medicine or psychology can be left to religion.

Occasionally, it is suggested that, within psychology, perhaps the NDE is the province of parapsychology. But in the last fifty years, parapsychology has been ill equipped to search for a concept of death and near-death linked to ideas of context and meaning. This is because, for most of its recent history, parapsychology has largely surrendered to experimental and survey work. The conceptual tools in neighboring social psychology, which might have enabled parapsychology to break free of such narrow traditions, were rarely if ever used.

Another reason for the lack of interest in the broader meanings of death that the NDE might suggest is that the NDE is commonly identified with religious concerns or issues. Many scientists believe that scientific meanings of death are constructed for different, impartial purposes. Faith, philosophy, and otherworld imagery are for religion and have no place in science.

These distinctions must seem quaint to some; more important, they are both untrue and misleading. They are untrue because scientific thought does indeed involve issues of faith, philosophy, and otherworld imagery. They are misleading because not recognizing the similarities between science and religion encourages the view that the two enterprises are opposites instead of rivals and competitors.

Since NDEs revive religious imagery of the afterlife, religious issues in moral philosophy, and religious conceptions of the dualist nature of self (body *and* soul), they are seen by some scientists as outright competitors for scientific offerings. The scientific project embraces philosophies of empiricism, positivism, and a materialist conception of the self (body only). The philosophies of science discourage any attraction to worlds that cannot be seen (empiricism) or for which there are no mathematical proofs (positivism). Immortality should not be sought in dualist conceptions such as body/soul or brain/mind or machine/spirit. Instead, one should think of immortality in species terms through genetic theories or cosmological ones that picture a time when consciousness might be engineered away from its bodily origins. Consciousness might live disembodied in galaxies around black holes, dust clouds, or "e + e-" plasma.[10] So much for the absence of philosophy and otherworld imagery in science.

As for faith, a number of historians of science have noted the ruling presence of this quality in science. The English scholar Mary Midgley describes faith as the belief in and acceptance of a map, or broad vision of the world, that is taken for granted and allows one to order the facts. It is a way of organizing one's data about the world. In this way, faith is no less important to Catholics than to Marxists or to scientific theorists about the world.

For instance, academic scientists today mostly refuse on principle to consider or publish any research about the topics now viewed as spooky and grouped under the heading of parapsychology. In this case the reason is a frankly metaphysical objection to causation that cannot be explained by the laws of physics. This may be a good reason, but the point is that it operates *before* considering the evidence. It simply forbids all interest in a range of topics which equally intelligent scientists at the end of the nineteenth century found extremely interesting—because their metaphysic was different. At that time, too, scientists were willing to discuss the metaphysical issue itself, but today they have put it outside their frontiers. Similar things happen constantly even where there is no suspicion of metaphysics. For instance, the theory of continental drift was long dismissed as unscientific, and so for a time were James Lovelock's suggestions about damage to the ozone layer.[11]

In the above ways, science has demonstrated a vested interest in its critical response to religion and in offering itself to the broader culture as an alternative, an alternative married to a particular philosophy and a specific vision of the future for humanity. Phenomena that appear to be, prima facie, outside these philosophies or images are either ignored as irrelevant to its interests or, if amenable, are colonized by their explanations. Science does not interest itself in meanings of death that might be shared unless these originate within its own ranks *and* have their materialist seal of approval. Unlike the holy rollers of medieval Christendom, the high priests of science offer meanings of death that they formulate themselves, outside the orbit of personal experience. In this way, the scientific search for the meaning of death is still being looked for in the autopsy rooms of hospitals instead of in the halls outside it. The reluctance to link context with experience in the NDE is part of this style of professional practice and philosophy.

Perhaps another reason for the lack of interest in context by neuroscience workers may have to do with the inherent bias that characterizes the modern—that is, modernist—scientific paradigm that underpins much of their work. Modern science has been following an increasingly narrow and decontextualizing program since it began in the Renaissance some 300 years ago. Stephen Toulmin traces these developments in the following way.[12]

Contrary to popular belief, the Renaissance gave birth not to one, but two, rational-scientific perspectives. The first, or *literary-humanist,* phase emerged amid the economic prosperity of the sixteenth century and celebrated human diversity. Its arguments privileged both logic and rhetoric, realizing that theoretical arguments had currency only on the basis of specific case examinations and audience characteristics. Logic was tied to the situation and context of a particular case, and its acceptance by the listener was contingent on how well its advocate was able to show how such a case related to the experience of the listeners.

This style of humanism discouraged intellectual dogmatism. It was thought that the pursuit of universal and absolute theories would encourage dogmatism, which, in religious contexts, can (and did) lead to major political life-and-death disputes. There was an avowed respect for the limits of both reason and experience, with emphasis on experience. As Toulmin wryly observes:

> Tolerating the resulting plurality, ambiguity, or the lack of certainty is no error, let alone a sin. Honest reflection shows that it is part of the price that we inevitably pay for being human beings, and not gods.[13]

The second scientific and philosophical phase of the Renaissance—the *rational-scientific* one—developed from the economic crises of the seventeenth century and led many thinkers to turn their backs on the first phase. Beginning with Descartes, this group of scholars attempted to develop universal theories independent of context. This concern for, and pursuit of, certainty was a response to the historical chaos and religious-political uncertainty of the times. The so-called project of modernity, then, was based on the establishment of a permanent, authoritative set of principles that would hold true everywhere, preferably irrespective of context. If the literary-humanist perspective studied the whole case in order to locate it in the particular scenario, the rational-scientific one paid less attention to the whole case in order to focus on a particular aspect of it. Rational scrutiny of parts of a whole, it was believed, would open up the secrets of that whole and then provide objective sources of guidance for future policy and action. In this way, the hopes of both modernity and of science became one.

The first child of the Renaissance, then, the literary-humanist period, is the origin of modern-day humanities—ethnography, literary studies, poetry, and history—all sciences in which context is integral. The second development of the Renaissance, the one that has dominated the modern imagination, is the parent of modern-day sciences—empirical philosophy, mathematics, physics, and biology.[14]

Each of these disciplines constructed its own set of narratives about the world, as well as its own internal debates about territorial borders. In this way, the development of academic disciplines paralleled the development of modern nation-states on issues of similar concern—cultural identity and political sovereignty.[15] The historical emphasis of both types of development was on *differences* rather than *shared* concerns and goals.

The bereavement narrative discussed earlier is a good example. As I mentioned, each discipline, whether from social work, parapsychology, or medicine, emphasizes *aspects* of a phenomenon that interests its readership. In this way, some aspects of experience are deliberately or inadvertently privileged or omitted. If the academic narrative is powerful in the community (e.g., medicine), its narrative will be better known than less influential narratives discussing the same subject matter (e.g., parapsychology). This is one reason why, for example, the visionary aspects of bereavement are a surprise to many people; they are simply not widely known compared to, say, Kubler-Ross's stage theory of grief.[16]

In any case, because so much of our understanding of the NDE, or bereavement, comes from the sciences rather than the humanities, there has been some methodological bias in favor of studying responses rather than experiences. Much of our knowledge in these areas has focused on the *char-*

acter of emotions or physical responses rather than on the specific experiences or circumstances that give rise to them.

I am not suggesting that the different emphases and omissions are due to an immoral conspiracy or a sense of rivalry. Rather, the separate concerns of each discipline as they have developed over the centuries have made these disciplines *confine* their respective interests and perspectives. Obviously, there have been important gains from these specializations and their respective intellectual habits. Nevertheless, it is not all good news.

One of the results has been to fragment the task of balancing a social portrayal of experience by delegating this responsibility to different disciplines. And, unless there is a deliberate attempt to cooperate in an area, a picture emerges that may only reflect the experience as understood by the advocates of a particular discipline; therefore, the picture is incomplete. What finally emerges from the academy as an "understanding" can more often be viewed as an artifact of the preferred portrayal by a popular or dominant academic perspective.

In summary, then, Toulmin's historical discussion of the emergence of disciplines has two further implications for the study of NDEs. First, both medicine and neuropsychology, disciplines that have modeled themselves on the physical sciences and their methods, tend to examine the NDE by breaking it into its components. Historically this is characteristic of sciences that have followed the rational-scientific development of the Renaissance. Here, in their search for a universal biological mechanism, analysis is atomized and narrowed. This process of breaking experiences up into discrete units for examination has enabled the taxonomic classification of microprocesses in biology and chemistry, among other sciences. However, a narrow search for mechanism can be misleading, particularly if the problem calls for an understanding of contexts (e.g., ecology, agricultural science, quantum mechanics, or near-death studies).

In current medical and psychological studies of the NDE, the study of trees has taken precedence over that of the forest. Every science is valuable in the study of NDEs, but those disinclined to link experience with organic process can miss the point. To continue the trees/forest analogy, knowing the *genetic* material responsible for plant growth will not predict that growth without knowing about growth *conditions*. It is soil, weather, predators, and competitors that hold the key to the appearance and survival of various plants, not simply the strength and adaptive nature of the genetic inheritance. Unfortunately, current medical and psychological studies of the NDE reflect these decontextualizing practices because their role models largely continue to be the basic physical sciences.

The second reason that this decontextualizing practice continues has to do with the lack of interdisciplinary work on the NDE. Academic disciplines since the Enlightenment have emphasized their unique and sovereign intellectual contributions rather than how their work is linked broadly with that of others. But for near-death studies, it is possible, indeed desirable, that social psychologists or medical anthropologists work with colleagues from neuroscience. This can lead to the development of explanatory theories about the NDE that are well grounded in the experiences of the percipient, balanced in their attention to the diversity of NDE or crisis phenomenology, and forged in the current thinking and debate of their respective disciplines.

Only by linking the current problems of consciousness and physical process to issues of identity and meaning and to those of tradition and culture can the study of the NDE make fruitful connections with other crisis experiences in life. This is also one of the few ways that the reductionism stemming from work in isolated disciplines can be checked and balanced by the critical considerations of others in the academy.

In conclusion, then, four major reasons appear to be responsible for taking the intellectual attempt to understand the NDE in a direction so different from that of the general community: the successful medicalization of NDE imagery in the academy; the academy's suspicion of things religious and its adherence to utopian images of science; scientists' preference for decontextualizing perspectives; and the fragmentation of the academic enterprise into isolated academic departments and disciplines. Each of these influences on the academy has led to poor responses to the NDE.

First, the academy has not acknowledged the links between the NDE and other crisis experiences that people undergo. Second, it has clearly underestimated both how common and how important these events are to most people. Third, these deficiencies in understanding have led large sections of the intellectual community to ignore the NDE, preferring to think of it, if they think of it at all, as someone else's problem. Of those who believe it is in their area of study, many believe that the NDE is a largely physical or psychoanalytic problem. This, in turn, has led to the development and stubborn persistence of a stereotyped and decontextualized set of images about the NDE. Finally, this has been disappointing for both NDErs and the community in general.

Many members of the community, confronted by this rather baffling response from its intellectuals, have turned to New Age, pop psychology, and/or Eastern religious literature. Those sources, and people's attraction to them, have been roundly criticized and lamented by advocates of scientific rationalism.

Nevertheless, as we have seen, such behavior is not pointless or based solely on successful marketing practices. These developments are responses—indeed, critical responses—to perceived inadequacies. Eastern religious philosophies seem attractive partly because Western religious philosophies seem less so for many people. Pop psychology books are pop not simply because they are easier to read or because they oversimplify. Another reason may be that, unlike the serious tomes of academic psychology, they address issues and events of genuine concern to ordinary people.

Furthermore, as Midgely observes,[17] since science deliberately competes with religion, it must therefore be competing for the same business. If the New Age is unscientific or pseudoscientific, perhaps real science has failed to convince ordinary people of science's merit. Real science may have disappointed in matters of ultimate importance to ordinary men and women, namely, the meaning of personal troubles in everyday life. If science does not address the personal and social problems of the people who have supported it, on what basis do its practitioners and supporters criticize those who look elsewhere for guidance?

The Future

To argue that OBEs, visions of the bereaved, or life reviews are part of the experience of crisis is by no means to settle the debate over explanations. Life review and OBEs do occur outside of crisis experiences. They also occur in reminiscences of the aged, and as part of psychotic and drug-induced experiences. The precise triggers, and their biological mechanisms, may or may not be the same. The question of their occasion is complex and goes beyond my preliminary introduction and discussion of them.

But the key to addressing these uncertainties is not to retreat from their complexity. A search for some neurotransmitter or psychoanalytic axiom that might tie these down to a single proposition is both inappropriate and inadequate to the task. The difficult challenge to serious intellectual students of the NDE is to create ways of thinking that are capable of including such diversities and anomalies.

Let us not forget that academic narratives (theories) are intellectual stories that attempt to bring seemingly *disparate* elements of experience *together*. We regularly entertain an assortment of theories, judging each on how many and how well it explains the diverse elements and relationships. In this scenario, the most specific, least complex phenomena with the least possibility of variance or change lead to the most stable theories. In this way, there is a pressure on all thinking (and thinkers) to oversimplify the phenomena under investigation. And although this is largely an unavoidable habit of the human mind in search of certainty, some topics attract more pressure than others. We have seen how certain historical and cultural elements in the academy work in this way for the study of NDEs.

But the most important purpose of introducing the idea of NDE as a crisis is not that this framework is more receptive to the complexities of the experience, though indeed it is. The most important reason is that crisis theory is broad enough in its characteristics to act as a net for other experiences that are clearly related. That method in the theory can alert students of the NDE to aspects of their particular phenomenon that are not readily apparent in the phenomenon itself. The social and environmental features of context are one of those important aspects.

Linking the NDE to other types of social experiences also draws the scientific gaze away from questions about death—and beyond death. On both historical and theoretical grounds, science has performed badly here, engaging in pointless argument, poorly equipped, in an unwinnable debate. Those who have lost their minds in this debate continue to push back the borders of argument and "proof" endlessly. As Zaleski argues, "We have seen enough, in any case, to guess that the final outcome will be a draw."[18] Why, then, except for reasons of ideological bigotry, should anyone continue on in this direction?

Not for nothing, then, is it more meaningful to return to the NDE as we know it and ask, as C. Wright Mills asked in a similar fashion some thirty-five years ago:[19] how can the personal crises of ordinary people be linked to their cultural traditions and biographical circumstances, and explained in terms of their intersections within psyche and organism? For the NDE as personal experience and as scholarly study, this is the only question with a genuine future.

APPENDIX: ON THEORY
AND METHOD

When presenting my work on the NDE to both academic and nonacademic audiences, I am regularly asked the same three or four questions. I have no doubt that a reading of this book will, among other things, stimulate some of these same questions in the reader. In an effort to clarify my approach, I have decided to summarize most of the common questions here and to provide a reasonably formal reply to them.

Sociological examination of the NDE has been unusual to date, given that the field of near-death studies has been dominated largely by clinical and philosophical perspectives. I hope some of the following questions and their answers may locate the intellectual place and social value of sociology in the study of crises such as those near death.

1. Why is a sociological perspective important to our understanding of the NDE?

This question is mostly asked by nonsociologists, so I will introduce my answer by explaining the importance of the sociological perspective in general.

Sociology was developed to understand the rapid, disorienting social changes occurring during the nineteenth century, particularly in Europe. Psychological explanations that attributed these changes to personal character were increasingly viewed as remote and abstract. Whether these changes could be attributed to the brutal nature or warlike temper of humanity did not explain why certain developments happened and why some of these occurred when they did and not, for example, earlier rather than later, in Germany rather than in England. In that context, sociology began to argue that human beings behave in certain ways for at least two reasons.

First, people are motivated by personal or private reasons. In the private world of the individual, anger, sorrow, guilt, or glee might make one person do one thing and another person act differently. The second influence on people's behavior is the social or public forces in their lives. In the public experience of life, people behave in many diverse ways for organizational reasons. Their occupation or public expectations demand that people act in preferred ways. This organizational aspect of life, this public experience of self, is commonly overlooked. People tend to miss the fact that much of their behavior is publicly expected rather than privately, spontaneously, individualistically designed and expressed. When people think of themselves, they think of their *differences* from others and overlook the fact that as bus conductors or social workers, much of their daily behavior resembles that of thousands of others who are also bus conductors or social workers.

Furthermore, much of our thinking and personal values are heavily influenced by social experiences. These social experiences, in turn depend on who we know. If you've been a bus conductor for thirty-six years, you will know fewer academics and politicians than an academic person will. You will also know even fewer homeless people. If you're an Australian working in Sydney, you will meet fewer Japanese people than if you're an Australian working in Tokyo. If you were born and raised in Kunming, China, the chances are that, for you, Christianity is a strange foreign religion. How well these insights are understood and integrated into any scholarly or casual perspective is a measure of what in sociology is called the *sociological imagination*. Context is not simply the background to a situation but rather the essential key to its meaning. In this connection, then, the sociological perspective is important to the study of the NDE in both specific and general senses.

To provide a context to the NDE as a private experience, the sociological perspective shows how different social experiences of dying (e.g., as a surgical patient, suicide, or accident victim) may influence the character of the NDE.

It also shows that NDEs are not simply medical or religious events but both and more. They are experiences at the edge of social regulation and control. In that respect, they are the accidental psychological consequences of isolation and marginalization. This is why tunnel sensation, life review, adopting an outsider's view of the self, and meeting deceased or supernatural beings as singular or combined experiences are not unique to the NDE. This is why many of these features occur to those contemplating suicide, to those in bereavement, to hapless shipwrecked castaways, and to people in mountaineering or mining accidents. It is also why these experiences are shared by those who deliberately take themselves to the edge of social experience—the shamans, the LSD users, the deep meditators, and the ascetics and mystics of both East and Western religions.

Furthermore, a sociological perspective is able to show that the content of many of these experiences is influenced by cultural matters, but not in the usual ways in which we have previously thought about them—gender, age, religious denomination, and so on. Rather, it seems that the religious ideas that dominate whole societies create the symbols and metaphors by which we communicate our stories about death, transition, suffering, and identity. The differences that are important comparatively are not those between countries but rather historical and anthropological differences in cosmological outlook.

Finally, a sociological perspective is important to provide a context to the NDE as a public experience. It shows that the widespread attraction to the NDE story is not about, or not simply about, the desire for immortality. NDEs and other stories about afterlife scenarios have been with us for a long time. The popularity now, the attraction to this particular image of death, are related to broader political and historical developments beginning early in this century and leading up to the postindustrial, postmodern period. Some of this encompasses the search for utopian ideas in all major areas of cultural life. But in addition, the disenchantment with science and medicine as a general discourse is rooted in the problems of immediacy or relevance in matters to do with death.

The sociological perspective is also important in showing that all writing, all knowledge, is a social activity; this includes scientific writing despite its rhetoric to the contrary. So much academic knowledge involves the simple process of comparing new experiences with old ones. We can see this in neuroscience's attempt to include the NDE in its bag of hallucinations, dreams, and temporal lobe seizure activity. We can also see this in sociology's attempt to include the NDE in its own bag of metaphors and parallels. This is not necessarily wrong, as long as we appreciate the advantages and limitations that such a methodology confers on its users and as long as we are conscious of using this methodology. In these cases, we must take the similarities seriously while at the same time respecting the differences. If the differences are glossed over or subsumed under other similarities, we risk the charge of reductionism.

2. By including the NDE with experiences of shipwreck or mountaineering falls, aren't you being reductionist? For example, are you not foreclosing on the possibility of life after death?

I have not foreclosed on the possibility of life after death (or on neuroscientific explanations, for that matter) because my own sociological explanations are not reductionist ones. I have always maintained that sociology is a social science, not an omniscience. The question of life after death is one for religion, philosophy, and psychic research, and it is unfair to criticize sociology for not being one of these disciplines or for not sharing their interest in this question. That question has never been one of the declared objects of study for sociology, and hence there are no conceptual or methodological tools to which we might look to for an answer to it.

I have also never attempted to rival biological theories of explanation. The only challenge suggested by my analysis is that NDEs may be triggered or prepared by social situations. These, in turn, give rise to psychological processes. In addition, these psychological reactions must have physiological correlates, as surely as does all of our cognitive-affective activity. In this sense, then, the biological mechanisms do not cause the NDE but rather are mediated by them.

Similarly, a smile, which has a physiological mechanism is not actually caused by that mechanism. Rather, the smile is a *response* to a social thought or circumstance; and the physiological mechanism responds to these, and not the reverse. This view of physiology does not minimize the importance of advances in neuroscience any more than it challenges the reality of the NDE as a possible glimpse of the afterlife. Let's return to our smile analogy once more.

Some people may smile because they are reacting to a private thought. Others smile with hundreds of others at a cinema viewing of a film. Still others smile at a three-dimensional experimental picture that others cannot see until they alter the way they usually see. What is the "real" content of these three examples—the thought that no one can see, the film that everyone can see, or the picture image that only some can see under certain perceptual conditions?

In each of these cases, the physiological mechanisms may be similar or different. This is a job for the neurophysiologists and neuropsychiatrists. What is "real" about thoughts, or film, or visual puzzles is something for the philosophers, social psychologists, and social theorists to consider. These are social and ontological questions. But let me emphasize that all of these examples have a social dimension. What's in the thought, the film, and the solving of the puzzle that makes people smile? The social dimension, then, may explain the smile in terms of the conditions that govern its onset and possibly some of its characteristics (maniacal, sympathetic, open-mouthed relief, etc.). But these explanations do not address the physiological or philosophical questions.

One of my tasks has been to explain the social conditions under which an NDE is most likely to occur and also how some of these conditions might promote some characteristics of the NDE over others. When unusual sociological phenomena have been associated with the NDE—enormous community interest or overreaction by neuroscientists—it has also been interesting to examine the cultural reasons behind this.

Finally, when I compare the mountaineer's fall with the castaway sailor, and then compare both of these to the person experiencing an NDE, I am not claiming that these are the *same* experiences. I am only arguing that these phenomena belong to a similar broad class of experiences. Let me employ another analogy. Imagine that I am studying the social experience of travel. In this case, I am not attempting to assess whether or not the descriptions of travelers match the actual places to which they claim to have traveled. Nor am I saying that traveling to Paris is the same as traveling to a town 50 miles from home. Nor am I saying that traveling for business is the same as traveling for pleasure. Finally, I am not saying that taking a trip to Europe is the same as taking a trip on LSD.

However, I am arguing that there are common, simple, and important elements in all of these experiences that warrant the label *traveling experiences*. For a sociologist, the question of the reality of any of the destinations is of secondary importance. Simply appearing in Paris does not guarantee a wonderful experience, any more than does appearing in Dubbo, Australia, guarantee an indifferent one. Better predictors of the personal experience of the trip are the expectations and experience of the traveler, the circumstances and resources available during travel, and the description of the trip itself by the traveler.

All of these may have little resemblance to or connection with the actual places or experiences as we know them, but that only points up the fact that all the foregoing *conditions* were different for *us*. One may also argue that going on an LSD trip is not really going away, and that traveling to Wagga Wagga is not the same as a trip to Moscow, but these are comments about quality rather than substance. And they may not even be major comments about quality. Whether a trip to Wagga Wagga can be compared to a trip to Moscow depends, first, on whether you come *from* Moscow or Wagga Wagga. Once again, it is expectations, purpose, prior experience, personality, and financial and social conditions that determine the comparison in major ways, not the physical location alone. The above comments notwithstanding, these are all still trips, experiences of travel.[1]

And NDEs are like the experiences of castaways and fallen mountaineers. Yes, there are differences between these experiences, but the differences are not great. Do castaways really see angels, as some NDErs claim to see? Possibly, possibly not. The social and psychoneurological conditions may make one situation ripe for such an encounter, and they may create the ideal material conditions for hallucinations that resemble these in yet another case.

I realize that it is common—and I believe erroneous—practice for scholars, particularly scientists such as Albert Einstein, Richard Dawkins, or Stephen Hawking, to pronounce on questions about God and other major human verities. However, there is no reason to believe that the folksy scientific rationalism of these scholars has greater merit than contemporary philosophical and theological debate. Not surprisingly, the latter groups have a methodological and theoretical preoccupation with questions about God, the afterlife, the mind/body problem, and so on, and bring a certain thoughtfulness and thoroughness to these questions. Given this epistemological set of circumstances, I have never entered a debate that is both unnecessary to the task of my analysis or one in which, in any case, I would be entirely ill equipped.

3. What place, then, does the NDE occupy in the sociological discourse? This question is more commonly asked by sociologists and anthropologists. It is more than simply a matter of neat housekeeping within the categories of sociology's specialties. Specialties, like whole disciplines, bring "ways of seeing" to phenomena that they select to study. Having said this, I will add that few topics or subjects of study occupy only one location within a discipline. Consider child abuse and prostitution. We should note that these experiences occupy important places of discussion in the sociology of the family, the sociology of health, the sociology of welfare, and the sociology of deviance, to name only the main perspectives.

In theory, the NDE may be part of the concern of the sociology of health, religion, and death and dying. As part of the sociology of health, the NDE is explored in terms of its effects on the experient and also of the medical response to it, both in social and in academic terms. As part of the sociology of religion, the NDE may be discussed in terms of its appropriation by New Age religious movements. It may also be discussed in terms of its place in rival eschatological doctrines.

Finally, the theological and political responses of established religions to the NDE, as well as the social and political links of the NDE with Eastern religions and philosophies, may also be encompassed by a sociology of religion. And of course, the sociology of death and dying may examine the NDE for what it may reveal about cultural behavior and expectations surrounding the circumstances of death and dying. In teaching, I confess to discussing the NDE mainly in the context of the sociology of health and the sociology of death and dying. However, I suspect that these are not the most appropriate forums for its examination. As previously noted, specialties, like disciplines, "see" the phenomena they select to study in particular ways. In this sense, I am less and less sure that the NDE should be seen as part of a medical/health concern or as a death-and-dying issue. I readily concede that several aspects of the NDE are relevant to these ways of seeing, but the deeper significance of the NDE may lie in their study as social deviance and marginality.

It seems clear from a study of shamanism, UFO abductions, shipwrecked castaways, and the experience of religious ascetics that the *basic* features of the NDE have little to do with death. What close brushes with death seem to have in common with being lost at sea or with abduction is the dramatic and traumatizing sense of social alienation and isolation.

The usual and expected experiences and associations are lost, however fleetingly, from the experient's point of view, perhaps forever—the company of friends and lovers; the taken-for-granted familiarity of meals, environments, sense of security and control habits of leisure, colors, smells, and sounds. The predicability of one's usual environment, combined with a sense of control over that environment, contributes to a feeling of safety. As the environment becomes more threatening, as with the increasing possibility of unemployment or a rising workload, a sense of alienation may occur and express itself as stress. In severe environmental disturbance such as the destruction of one's usual environment through disaster (bush fires, earthquake, war, etc.), the disordered experience can lead to revision of many taken-for-granted values and attitudes, as well as unusual thoughts and emotions in the midst of the disaster experience itself. Unusual events bring unusual personal experiences.

Correspondingly, if the environment remains stable but one's sense of control slips, as in psychosis, severe anxiety, or depression, the result is similar for the individual. A small drop in the personal sense of control can induce feelings of vulnerability and isolation. This can often lead to deepening feelings of the initial anxiety or depression. With complete psychological inadequacy, the depressed or psychotic person may be institutionalized, completing, his or her feelings of vul-

nerability and isolation, now joined by other feelings such as depersonalization and stigma.

However the balance is tipped, people regularly experience a sense of
marginalization from what is seen, correctly or incorrectly, as the normal
working of life itself. A bout of unemployment or a severe illness, the experiences of war, or a traumatizing divorce or accident can give many people
pause. Life is never the same for many of these people, with the social
experiences evoking thoughts and feelings they never thought they had or
could have; these are events and situations that they thought happened only
to other people.

Some experiences are even more disorientating, such as being lost at sea,
being widowed, or undergoing shamanic initiation. These seem to generate
even more profound and dramatic psychological experiences. The persons
involved may see deceased friends or relatives or gods. They may experience
euphoria or fear of a depth previously unknown. Such circumstances are
related directly to the experience of marginality; the perception of being
socially marginal and isolated by the experience seems to prompt them. Erikson has said:

> Deviance is not a property inherent in any particular kind of behavior; it is a
> property conferred upon that behavior by the people who come into direct or
> indirect contact with it.[2]

This understanding of marginality is a common one in the social sciences,
even a dominant one. The problem with this view is that it is purely behav-
ioral, not acknowledging the *experience* of deviance from the point of view
of the experiencer. In that respect, it emphasizes the social labeling of the
problem of deviance. The focus is on the status of deviance—how it is
expressed, labeled, and responded to by others.

However, people not only identify deviance, they also feel and experience
it themselves. The castaway feels lost; the NDEr feels cut off from others—
sometimes before the brush with death, sometimes during the OBE. Betty
Eadie, in her best-selling book *Embraced by the Light*, recounts a common
feeling of some NDErs:

> But when I hung up [the phone] the loneliness fell on me again like a blanket.
> The room seemed darker, and the distance between the hospital and our home
> felt more like a million miles than just across town. My family was life itself
> to me, and being away from them scared me, hurt me.[3]

This sense of isolation, of separation from cherished people and experiences, may be responsible for new social and psychological experiences, which in turn can lead to new attitudes and behaviors. And even though these new thoughts, values, and behaviors appear coherent and normal to the experient, others may not share this perception. At that point, the experient may be accorded a positive response by the community, as received by shamans or heroic survivors of ordeals. Alternatively, they may experience stigma or rejection, as do prostitutes, some NDErs, or the mentally ill. None of these groups may consider themselves deviant in any criminal or moral way. Their marginal status is derived from undergoing unusual experiences that have promoted new ways of thinking and feeling for them, and these, in turn, have attracted new social responses to them. The initial feeling of isolation may linger and become a permanent part of their social and psychological response; this, in turn, may lead others to treating them accordingly.

Precisely what social conditions prompt these unusual experiences, and how they create different social and emotional changes in the individual, is an important part of the study of the sociology of deviance and marginality. I therefore feel that, perhaps this area of sociology may further our sociological understanding of the NDE.

4. Patricia Weibust,[4] in commenting on your analysis of images of utopia in the NDE, argues that most of your sociological material was drawn from Craig Lundahl's study of Mormon NDEs: "who . . . besides the nine mormons saw these particular sociological phenomena?"

I couldn't understand why Weibust made this claim, since in my original article I made specific reference to other supporting literature. Craig Lundahl's work was privileged not because he employed Mormon accounts but because these accounts contained particular detail of a transcendent society. Images of society in the NDE, however, are also contained in the work of Moody, Ring, Pasricha and Stevenson, and Counts and Elder—all of whom are cited in the original article.

However, extended accounts with significant corroborating social and political detail in NDEs are available in individual recollections of the NDE. These can be compared with my description of the transcendent society by anyone who cares to take the time. Of particular relevance here are the accounts of Elder,[5] Ring,[6] Ritchie,[7] Black Elk/Neihardt,[8] and, most recently, Eadie.[9]

5. Antonia Mills, also commenting on your attempts to sketch a theory of society from elaborate NDEs, feels that the task is somewhat pointless:

One would not expect travelers who only came into foreign airports or landing strips, were met by a welcoming committee, and then sent back, to tell us much about the qualities and characteristics of the societies behind the airports and landing strips. The same applies to NDEs.[10]

I agree with Mills's analogy of people merely visiting airports. However, the NDE accounts used by me were not of the simple airport greeting kind. Such accounts *do not* contain societal description. That is the whole point. The accounts used by me were drawn from the exceptional accounts of NDErs who were privileged to travel *beyond* the "airport" or "landing strip." These people went with their "welcoming party" on a "quick tour of town."

Consider some of George Ritchie's narrative hinting at the presence of education, work, and services in the "other world":

> But the atmosphere of the place was not at all as I imagined a monastery. It was more like some tremendous study center, humming with the excitement of great discovery. Everyone we passed seemed caught up in some all engrossing activity; . . . Through open doors I glimpsed enormous rooms filled with complex equipment. In several of the rooms hooded figures bent over intricate charts and diagrams, or sat at the controls of elaborate consoles flickering with lights.[11]

> Next we walked through a library the size of the whole of the University of Richmond. I gazed into rooms lined floor to ceiling with documents on parchment, clay, leather, metal, paper.[12]

> And then I saw, infinitely far off, far too distant to be visible with any kind of sight I knew of . . . a city.[13]

Bettie Eadie's recent account of her own prolonged NDE confirms other sociological features discussed by me when describing and examining the transcendent society. One is the high level of social organization displayed there. Eadie describes this social world as governed by "laws."[14]

> I saw that there are many laws by which we are governed—spiritual laws, physical laws, universal laws—most of which we have only an inkling.[15]

Another chapter in her account describes a library, as well as large loom and its workers.

The workers explained that the material would be made into clothing for those coming into the spirit world from Earth.[16]

Clearly, these are not "airport" accounts; they are also not unprecedented. Hints of this society and its organization appear in Raymond Moodys early work.
One NDEr in Moody's book recounts:

Around the edges of the door I could see a really brilliant light, with rays just streaming like everybody was so happy in there, and reeling around, moving around. It seemed like it was awfully busy in there.[17]

Beyond the mist, I could see people, and their forms were just like they are on earth, and I could also see something which one could take to be buildings.[18]

Of the importance of learning, even in the transcendent society, another NDEr says:

He seemed very interested in things concerning knowledge, too. . . . He said that it is a continuous process, so I got the feeling that it goes on after death.[19]

I have not made large, complex claims for my description of the transcendent society. But the description drawn from these and other accounts is substantial and consistent enough to claim that it is *organization* that is central to the transcendent society. Love, service/work, and learning in structured social environments are important features. The control of deviance is also present. These features allow the society at the center of these NDEs to be described as utopian rather than as divine rule/millenarian or any other type of ideal society. And that, as I have argued, has important implications for the way we understand its political and cultural significance in modern urban society.

6. A question concerning your general methodology: Most of your analysis is based on secondary analysis. How can you really understand the NDE unless you actually interview or talk to those who have undergone the experience?
Let me begin my answer with some general remarks before focusing on NDE research in particular. I began an earlier book, entitled *The Unobtrusive Researcher*, by making the following observation:

There is today, in social sciences circles, a simple and persistent belief that knowledge about people is available simply by asking. We ask people about themselves and they tell us. Either we ask them a series of questions in a survey or we have a discussion with them in a structured, semi-structured or "unstructured" interview. In any case, the assumption is that important "truths" about people are best gained through talk.[20]

I went on from there to point out all the other ways that one might gain insight into the reasons why people do and experience things. There is no one "right" way to do social research, nor is there one consistently superior method. If this were not so, historians, archaeologists, and economists would be out of work because they rarely employ interviews. Alternatively, psychologists and sociologists would be out of work because they so often rely on interviews. Certain methods are regularly used by one group because they best serve the research questions being asked. Methods and questions go together. Methods, in themselves, are no measure of the value or quality of any piece of social research.

Interviews are valuable but, like all methods, not infallible. People do not swallow a "truth serum" when being interviewed. Furthermore, even honest and open interviews have their limits. This is because not everyone knows why they act, think or feel, or believe as they do. Asking them does not always help. In this case, interview data are not more valid than data from secondary analysis. There are two reasons for this.

First, there are imperfections of recall that result in selectivity of memory; there are problems of interviewer bias; and there are problems of interviewer relationships that can result in interviewees responding to an expectation, or resisting that expectation, from interviewers. In this instance, there are further complications arising from problems of trust, rapport, and miscommunication—the usual problems that dog all social encounters with anyone, let alone strangers. So, firsthand, collected interview data are no more a source of "clear" and "unpolluted" information than existing records or accounts.

Second, once the data are collected, however great their integrity, the problem of interpretation begins. Simply collecting the facts does not convey an experience. Telling the experience to others requires the story to be organized. This process begins with the interviewee but is aided by the interviewer, who asks the questions or guides the account by supplying topics or empathy and encouragement. Once the account is collected, the researcher may compare it with his or her own theory (the hypothetico-deductive approach) or search the account thematically (the ethnographic-inductive approach). No search for themes in other people's accounts is value neutral.

Themes that "emerge" do so as a function of the social relationship between the listener and the teller, the reader and the text. One brings to a reading one's own past experiences, sensitivities, and blind spots. So, interpreting interview data has its own hazards in terms of the reliability and validity of the data. In summary, then, interviewing is not superior to other research methods.

These reservations and qualifications notwithstanding, interviews in the near-death study area are valuable for the insights they have given us, particularly in the early years of research in this area. It was Moody's, Ring's, and Sabom's in-depth interviews that gave us the first full phenomenological accounts of the experience. More recently, Greyson and Bush[21] have continued to use this method in their work with negative NDEs. For consistently sharp details of the NDE as a personal experience, the interview has been an excellent revelatory tool.

However, the last few years have seen prominent workers employ the interview with little new insight. As one journalist recently asked me, "What, aside from OBEs, tunnels, white lights, life review and personal transformation, have we learned about the NDE since Ken Ring wrote in 1980?" To answer this question, I pointed out Carol Zaleski's *Otherworld Journeys*,[22] which compared medieval NDEs with contemporary accounts. David Lorimer's *Whole in One*[23] was also a milestone in examining the NDE for its possible philosophical role as the basis of a new ethics. But what is striking in these works is that they are based on two important methodological strategies. First, they use existing sources—available NDE accounts and other academic and professional literature. Second, they have moved away from the simple phenomenological account and attempted to develop theoretical explanations and questions that might advance religious studies or philosophical understanding of the NDE. Recently, Susan Blackmore has adopted a similar approach in her neuroscientific work on the NDE.[24] For the first ten years or so, it was important to ensure that the NDE was well described. But the larger task confronting those of us in religious studies, medicine, or sociology in the twenty-first century is to elucidate the *meaning* of the NDE. That task requires different theoretical and methodological approaches to the topic, and this has also been the rationale behind the broad shape and style of my own sociological work in this area at this time.

7. You argue in Chapter 2 that the tunnel experience does not occur in India or in hunter-gatherer societies, but Blackmore observed tunnels in her Indian study of the NDE.[25]

I have not argued that the tunnel sensation is not part of the Indian NDE.

Rather, I have observed that it has apparently not been reported by NDE researchers in that country. When I originally wrote my survey, I relied principally on the work of Osis and Haraldsson[26] and Pasricha and Stevenson.[27] Subsequent work by Pasricha in India has also not reported the tunnel sensation.[28] Between the time of the Pasricha and Stevenson study and the later ones conducted by Pasricha, the number of Indian cases has totaled forty-five.

Blackmore's study and conclusions were based on a sample of eight reports, only one of which mentions the tunnel sensation. With Stevenson and his colleagues, I reject Blackmore's conclusions.[29] Let me briefly summarize our reasons. First, only three Indians reported an experience of "great darkness" in their written correspondence with Blackmore. These three correspondents were then sent a questionnaire, which asked the leading question "Did you see a tunnel?" One accepted this description. However, another respondent may have accepted another description—for example, a cave, a dark room, and so on. The only valid, reliable method is to allow people to choose their own description for their experiences.

But even allowing for the possibility that Blackmore, or anyone else, receives a report of a tunnel sensation from India, this would not be surprising. For as I said in Chapter 2, people from societies dominated by historical religions will be more likely to report tunnel sensations than people from, for example, hunter-gatherer cultures. This is because societies with historical religions are traditionally long-term settlement cultures. These are societies where tunnels are well known technologically, architecturally, and intellectually. India, of course, is such a society; so, in theory at least, reports of a tunnel sensation may be anticipated.

Conversely, I have argued that life review and the tunnel sensation seem "confined *largely* to societies where historic religions are dominant." By this I meant that because darkness appears to be a cross-cultural experience, the descriptor "tunnel" might be much more common in historical societies because the people in those societies have daily encounters with that shape. Those in hunter-gatherer societies tend not to encounter this symbol regularly, so this would lead us to assume that this descriptor will be rarely chosen: not *never* chosen, but rarely. I have not been able to find the tunnel as a descriptor in any NDE account from societies dominated by archaic or primitive religions. And despite being open to the possibility that one *may* encounter such a description, I argue only two issues. First, darkness will rarely be described as tunnel-like. Second, such issues are best understood in terms of language rather than those of any so-called physiology. If every NDE description led a biological hunt for its cause, we would be developing physiological theories for cave sensations or the gardens often reportedly seen in these experiences! This would be taking metaphorical descriptions literally—a methodological tendency shared by some neuroscientific and New Age theorists. Such literal interpretations lead to the belief in an empirical world independent of the social meanings that have created it—a problematic habit well discussed by Zaleski.[30]

NOTES

CHAPTER 1

1. George Ritchie, *Return from Tomorrow* (Waco, Tex.: Chosen Books, 1978).

2. Betty Eadie, *Embraced by the Light* (Placerville, Calif.: Gold Leaf Press, 1992).

3. Susan Blackmore, *Dying to Live: Science and the Near-Death Experience* (London: Grafton, 1993).

4. Raymond Moody, *Life after Life* (Covington Ga.: Mockingbird, 1975).

5. Kenneth Ring, *Life at Death: A Scientific Investigation of the Near-Death Experience* (New York: Coward, McCann & Geohegan, 1980).

6. C. Wright Mills, *The Sociological Imagination* (New York: Oxford University Press, 1959).

7. Dorothy Counts, "Near-Death and Out-of-Body Experiences in a Melanesian Society," *Anabiosis* 3 (1983): 119–20.

8. Moody, *Life after Life*, pp. 23–24.

9. Bruce Greyson and Barbara Harris, "Clinical Approaches to the Near-Death Experience," *Journal of Near-Death Studies* 6 (1987): 42–43.

10. Gordon Allport, *The Nature of Prejudice* (New York: Doubleday, 1958).

11. Allan Kellehear, *Dying of Cancer: The Final Year of Life* (London: Harwood Academic Publishers, 1990), pp. 100–4.

12. Kenneth Ring, "Amazing Grace: The Near-Death Experience as Compensatory Gift," *Journal of Near-Death Studies* 10 (1991): 30–31.

13. J. C. Saavedra-Aguila and J. S. Gomez-Jeria, "A Neurobiological Model for Near-Death Experiences," *Journal of Near-Death Studies* 7 (1989): 210.

14. Norman O. Brown, *Life Against Death: The Psychoanalytic Meaning of History* (London: Sphere), 1968 p. 95.

15. See Ernest Becker, *The Denial of Death* (New York: Collier-Macmillan, 1973), and Allan Kellehear, "Are We a Death-Denying Society? A Sociological Review," *Social Science & Medicine* 18 (1984): 713–23.

16. See Oskar Pfister, "Shock Thoughts and Fantasies in Extreme Mortal Danger," translated by R. Kletti and R. Noyes, Jr., as "Mental States in Mortal Danger,"

Essence 5 (1981): 5–20. R. Noyes, Jr., "The Experience of Dying," *Psychiatry* 35 (1972): 174–83.

17. For a good review of the psychoanalytic work in this area, such as it is, see C. Zaleski, *Otherworld Journeys: Accounts of Near-Death Experiences in Medieval and Modern Times* (New York: Oxford University Press, 1987): 170–75.

18. See Carl Sagan, *Broca's Brain: Reflections on the Romance of Science* (New York: Random House, 1979).

19. Carl Becker, "The Failure of Saganomics: Why Birth Models Cannot Explain Near-Death Phenomena," *Anabiosis* 2 (1982): 102–9.

20. S. J. Blackmore, "Birth and the OBE: An Unhelpful Analogy," *Journal of the American Society for Psychical Research* 52 (1984): 225–44.

21. Margery Williams, *The Velveteen Rabbit: or How Toys Become Real* (New York: Doubleday and Co., 1922).

CHAPTER 2

1. See Moody, *Life after Life;* Ring, *Life at Death;* B. Greyson and I. Stevenson, "The Phenomenology of Near-Death Experiences," *American Journal of Psychiatry* 137 (1980): 1193–96; Michael Sabom, *Recollections of Death: A Medical Investigation* (New York: Harper & Row, 1982) for their work with American subjects. In England, see Margot Grey, *Return from Death: An Exploration of the Near-Death Experience* (London: Arkana, 1985). In Australia, see Cherie Sutherland, *Transformed by the Light* (Sydney: Bantam, 1992). The NDE has been contexualized in Western religious history by Zaleski, *Otherworld Journeys.*

2. See, for example, M. Grosso, "Toward an Explanation of Near-Death Phenomena," *Journal of the American Society for Psychical Research* 75 (1981): 37–60, and P. M. H. Atwater, *Coming Back to Life: The After-Effects of the Near-Death Experience* (New York: Ballantine Books, 1988).

3. S. J. Blackmore and T. S. Troscianko, "The Physiology of the Tunnel," *Journal of Near-Death Studies* 8 (1989): 15–28.

4. R. Noyes and R. Kletti, "Panoramic Memory: A Response to the Threat of Death," *Omega* 8 (1977): 181–94.

5. R. N. Butler, "The Life Review: An Integration of Reminiscence in the Aged," *Psychiatry* 26 (1963): 65–76.

6. C. R. Lundahl, "The Perceived Otherworld in Mormon Near-Death Experience: A Social and Physical Description," *Omega* 12 (1981–82): 319–27.

7. Bruce Greyson, "A Typology of the Near-Death Experience," *American Journal of Psychiatry* 142 (1988): 967–69.

8. Sabom, *Recollections of Death.*

9. Several workers have investigated NDEs in Western children. See, for example, N. E. Bush, "The Near-Death Experience in Children: Shades of the Prison House Re-Opening," *Anabiosis* 3 (1983): 177–93; Melvin Morse, D. Connor, and D. Tyler, "Near-Death Experiences in a Pediatric Population," *American Journal of*

Diseases of Children 139 (1985): 595–600; Melvin Morse with Paul Perry, *Closer to the Light* (London: Souvenir, 1990). The initial observation by Bush and Morse that children may not experience life review is qualified by other observers who note life reviews in children old enough to remember a social life. See Noyes and Kletti, "Panoramic Memory," and William Serdahely, "Pediatric Near-Death Experiences," *Journal of Near-Death Studies* 9 (1990): 33–39.

10. Carl Becker, "The Centrality of Near-Death Experiences in Chinese Pure Land Buddhism," *Anabiosis* 1 (1981): 154–71; also Carl Becker, "The Pure Land Re-Visited: Sino-Japanese Meditations and Near-Death Experiences of the Next World," *Anabiosis* 4 (1984): 51–68.

11. Feng Zhi-ying and Liu Jian-Xun, "Near-Death Experiences among Survivors of the 1976 Tangshan Earthquake" *Journal of Near-Death Studies* 11 (1992): 39–48.

12. Becker, "The Pure Land," p. 163.

13. Ibid.

14. S. Ogasawara, *Chugoku Kinsei Jodokyoshi no Kenkuy (Research on the History of Pure Land Buddhism in Recent China)* (Kyoto: Hyakkaeu, 1963).

15. W. Lai, "Tales of Rebirths and the Later Pure Land Tradition in China," in M. Solomon (ed.), *Berkeley Buddhist Studies* Series 3 (Berkeley, Calif.: Asian Humanities Press, in press).

16. K. Osis and E. Haraldsson, *At the Hour of Death* (New York: Avon, 1977).

17. Becker, "The Pure Land," p. 64.

18. Ibid, p. 59.

19. Feng Zhi-ying and Liu Jian-xun, "Near-Death Experiences."

20. Allan Kellehear, Patrick Heaven, and Jia Gao, "Community Attitudes Toward the Near-Death Experience: A Chinese Study," *Journal of Near-Death Studies* 8 (1990): 163–73.

21. Osis and Haraldsson, *Hour of Death.*

22. S. Pasricha and I. Stevenson, "Near-Death Experiences in India: A Preliminary Report," *Journal of Nervous and Mental Disease* 174 (1986): 165–70.

23. S. Pasricha, "A Systematic Survey of Near-Death Experiences in South India," *Journal of Scientific Exploration* 7 (1993): 161–71; and S. Pasricha, "Near-Death Experiences in South India: A Systematic Survey in Channapatua," *NIH-MANS Journal* 10 (1992): 111–18.

24. S. J. Blackmore, "Near-Death Experiences in India: They Have Tunnels Too," *Journal of Near-Death Studies* 11 (1993): 205–17.

25. For a comprehensive critique of Blackmore's Indian survey, see Allan Kellehear, Ian Stevenson, Satwant Pasricha, and Emily Cook, "The Absence of Tunnel Sensation in Near-Death Experiences from India," *Journal of Near-Death Studies* 13 (1994): 109–13.

26. J. T. Green, "Near-Death Experiences in a Chommorro Culture." *Vital Signs* 4 (1984): 1–2, 6–7.

27. Ibid, p. 6.

28. Counts, "Near-Death and Out-of-Body Experiences."

29. Ibid., p. 129.

30. Pasricha and Stevenson, "Near-Death Experiences."

31. Counts, "Near-Death and Out-of-Body Experiences." p. 123.

32. Ibid, p. 119.

33. C. E. Schorer, "Two Native North American Near-Death Experiences," *Omega* 16 (1985): 111–13.

34. H. R. Schoolcraft, *Travels in the Central Portion of the Mississippi Valley* (New York: Collins and Henry, 1825).

35. Juan S. Gomez-Jeria, "A Near-Death Experience among the Mapuche People," *Journal of Near-Death Studies* 11 (1993): 219–22.

36. Ibid., pp. 220–21.

37. Ibid, p. 221.

38. R. M. Berndt and C. H. Berndt, *The Speaking Land: Myth and Story in Aboriginal Australia* (Harmondsworth, U.K.: Penguin, 1989).

39. W. L. Warner, *A Black Civilization: A Social Study of an Australian tribe* (New York: Harper and Brothers, 1937).

40. Berndt and Berndt, *The Speaking Land,* p. 376.

41. Ibid, p. 381.

42. Ibid.

43. M. King, *Being Pakeha: An Encounter with New Zealand and the Maori Renaissance* (Auckland, N.Z.: Hodder & Stoughton, 1985).

44. Ibid., p. 92.

45. Ibid., pp. 93–94.

46. M. Panoff, "The Notion of the Double Self Among the Maenge," *Journal of the Polynesian Society* 77 (1968): 275–95.

47. D. Shiels, "A Cross-Cultural Study of Beliefs in Out-of-the-Body Experiences, Waking and Sleeping," *Journal of the Society for Psychical Research* 49 (1978): 697–741.

48. J. J. Hearney, *The Sacred and the Psychic* (New York: Paulist Press, 1984).

49. C. Levi-Strauss, *Totemism* (Harmondsworth, U.K.: Penguin, 1973).

50. Becker, "The Pure Land."

51. Blackmore and Troscianko, "Physiology of the Tunnel."

52. *Oxford English Dictionary* 18 (1989).

53. K. J. Drab, "The Tunnel Experience: Reality or Experience?" *Anabiosis* 1 (1981): 126–52.

54. Lewis Carroll, *Alice's Adventures in Wonderland* (New York: Airmont, 1965).

55. Gates do seem to play this major symbolic role of transition in Gomez-Jeria, "A Near Death Experience," cited earlier in the chapter.

56. Butler, "The Life Review."

57. Ibid, p. 75.

58. R. N. Bellah, *Beyond Belief: Essays on Religion in a Post-Traditional World* (New York: Harper and Row, 1976).

59. The term *primitive* is now enjoying a revival. No longer used in a moral or

evolutionary pejorative sense, *primitive,* as used here, refers only to the simpler and less differentiated social organizations. It also refers to a particular less differentiated relationship between self and environment. For a fuller discussion of these ideas, see M. Douglas, *Purity and Danger* (London: Routledge and Kegan Paul 1966).

60. G. Roheim, "Psychoanalysis of Primitive Cultural Types," *International Journal of Psychoanalysis* 13 (1, 2) (1932).

62. Ibid., p. 120.

63. M. Weber, *The Sociology of Religion* (London: Methuen, 1965), p. 43.

CHAPTER 3

1. Max Weber, *The Sociology of Religion.*

2. Emile Durkheim, *The Elementary Forms of the Religious Life* (New York: Free Press, 1965).

3. See Moody, *Life after Life;* Sabom, *Recollections of Death;* B. Greyson, "Toward a Psychological Explanation of Near-Death Experiences: A Response to Dr. Grosso's Paper," *Anabiosis* 1 (1981): 88–103.

4. See, for example, Ring, *Life at Death;* K. Drab, "Unresolved Problems in the Study of Near-Death Experiences: Some Suggestions for Research and Theory," *Anabiosis* 1 (1981): 27–43; Robert Kastenbaum, *Is There Life after Death?* (London: Prentice-Hall 1984); Grey, *Return from Death.*

5. R. K. Seigal, "The Psychology of Life after Death," *American Psychologist* 35 (1980): 911–31; R. K. Seigal, "Accounting for Afterlife Experiences," *Psychology Today* (Jan. 1981): 65–75; Blackmore, *Dying to Live.*

6. Noyes and Kletti, "Depersonalization."

7. C. P. Flynn, *After the Beyond: Human Transformation and the Near-Death Experience* (Englewood Cliffs, N.J.: Prentice-Hall, 1986), p. 9.

8. B. G. Glaser and A. L. Strauss, *Status Passage* (London: Routledge and Kegan Paul, 1971).

9. Ibid., p. 2.

10. B. G. Glaser and A. L. Strauss. *Awareness of Dying* (New York: Aldine, 1965); B. G. Glaser and A. L. Strauss, *Time for Dying* (Chicago: Aldine, 1968).

11. W. Y. Evans-Wentz, *The Tibetan Book of the Dead* (London: Oxford University Press, 1960).

12. Glaser and Strauss, *Status Passage,* p. 2.

13. See, for example, the survey by R. M. Veatch and E. Tai, "Talking about Death: Patterns of Lay and Professional Change," *American Academy of Political and Social Sciences* 447 (1980): 29–45.

14. Greyson and Harris, "Clinical Approaches."

15. D. Robertson, *Survive the Savage Sea* (London: Elek Books, 1973).

16. M. Bailey and M. Bailey, *Staying Alive!* (New York: David McKay, 1974).

17. "My Tale, by the Yacht's Skipper," *The Age* (7 Oct. 1989): 1, 4.

18. J. Culver, "Ordeal at Sea," *The Australian Magazine* (30 Sept.–1 Oct. 1989): 8–14.

19. Ibid., p. 13.

20. Robertson, *Survive*, p. 45.

21. The Age, p. 4.

22. Robertson, *Survive*, p. 216.

23. S. Simon-Buller, V. A. Christopherson, and R. A. Jones, "Correlates of Sensing the Presence of a Deceased Spouse," *Omega* 19 (1988–89): 21–30.

24. R. A. Kalish and D. Reynolds, "Phenomenological Reality and Postdeath Contact," *Journal for the Scientific Study of Religion* 12 (1973): 209–21.

25. P. Suedfield and J. S. P. Mocellin, "The 'Sensed Presence' in Unusual Environments," *Environment and Behavior* 19 (1987): 33–52.

26. The Age, p. 14.

27. Culver, "Ordeal at Sea," p. 14.

28. Robertson, *Survive*, p. 217.

29. Ring, *Life at Death*, p. 40.

30. See Greyson, "A Typology of Near-Death Experiences," and Ian Stevenson, Emily Cook, and N. McClean-Rice, "Are Persons Reporting 'Near-Death Experiences' Really Near Death? A Study of Medical Records," *Omega* 20 (1989–90): 45–54.

31. R. A. Lucas, "Social Implications of the Immediacy of Death," *Canadian Review of Sociology and Anthropology* 5 (1968): 1–16.

32. R. Noyes, "The Experience of Dying," *Psychiatry* 35 (1972): 174–84.

33. R. A. Kalish, "Non-Medical Interventions in Life and Death," *Social Science and Medicine* 4 (1970): 655–65.

34. Noyes, "Experience of Dying."

35. Greyson, "A Typology of Near-Death Experiences."

36. Noyes, "Experience of Dying," p. 179.

37. Lucas, "Social Implications," p. 11.

38. Noyes, "Experience of Dying," p. 182.

39. L. N. Tolstoy, *The Death of Ivan Ilyich* (New York: New American Library, 1960).

40. See, for example, Ring, *Life at Death;* Noyes, "Experience of Dying"; B. Greyson, "Near-Death Experiences and Attempted Suicides," *Suicide and Life Threatening Behavior* 2 (1981): 10–16.

41. Flynn, *After the Beyond.*

42. Ibid., p. 64.

43. Grey, *Return from Death.* However, also see B. Greyson and N. Evans-Bush, "Distressing Near-Death Experiences," *Psychiatry* 55 (1992): 95–110.

44. Greyson, "A Typology of Near-Death Experiences."

45. Lucas, "Social Implications."

46. R. A. Moody and P. Perry, *The Light Beyond* (London: Macmillan, 1988), p. 22.

47. Mills, *The Sociological Imagination.*

CHAPTER 4

1. Ring, *Life at Death*, p. 169.

2. Ibid.

3. Cherie Sutherland, "Changes in Religious Beliefs, Attitudes and Practices Following Near-Death Experiences: An Australian Study," *Journal of Near-Death Studies* 9 (1990): 28.

4. Leslee Morabito, "Love and God in the Near-Death Experience" (letter), *Journal of Near-Death Studies* 9 (1990): 65.

5. Atwater, *Coming Back to Life*, p. 89.

6. Flynn, *After the Beyond*, p. 31.

7. Bruce Elder, *And When I Die, Will I Be Dead?* (Sydney: Australian Broadcasting Commission, 1987), p. 41.

8. Kenneth Ring, *Heading Toward Omega: In search of the Meaning of the Near-Death Experience* (New York: William Morrow, 1984), p. 133.

9. George Gallup, Jr., with William Proctor, *Adventures in Immortality* (London: Souvenir Press, 1982), p. 130.

10. Ibid., p. 131.

11. Flynn, *After the Beyond*, p. 40.

12. Ring, *Heading Toward Omega*, p. 137.

13. Moody, *Life after Life*, p. 36.

14. Sutherland, "Changes in Religious Beliefs," p. 29.

15. Moody, *Life after Life*, pp. 94–95.

16. Ibid., p. 96.

17. Ibid., p. 87.

18. Ibid., pp. 84–88.

19. Mori Insinger, "The Impact of Near-Death Experience on Family Relationships," *Journal of Near-Death Studies* 9 (1991): 141–81.

20. Ibid., p. 150.

21. Ibid., p. 153.

22. Ibid., p. 159.

23. Ibid., p. 158.

24. Allan Kellehear and Patrick Heaven, "Community Attitudes Toward the Near-Death Experience: An Australian Study," *Journal of Near-Death Studies* 7 (1989): 165–72.

25. Kellehear, *Dying of Cancer: The Final Year of Life*.

26. Gallop, *Adventures in Immortality*, p. 183.

27. See Tony Walter, "Death in the New Age," *Religion* 23 (1993): 127–45. Also see the excellent and sometimes entertaining volume edited by R. Basil, *Not Necessarily the New Age: Critical Essays* (New York: Prometheus, 1989).

28. The most well-known of these groups is the Committee for Scientific Investigation of Claims of the Paranormal (CSICOP). It is a largely male organization

dedicated to a particular rather than a pluralist notion of science. For an instructive introduction to its history and social characteristics, see G. P. Hansen, "CSICOP and Skepticism: An Emerging Social Movement," *Journal of the American Society for Psychical Research* 86 (1992): 19–63.

29. Gallop, *Adventures in Immortality*, p. 207.

30. Insinger, "The Impact," p. 151.

31. Kellehear, Heaven, and Gao, "Community Attitudes."

32. Zaleski, *Otherworld Journeys*.

33. Hans Kung, *Eternal Life?* (London: Collins, 1984).

34. Ibid., p. 35.

35. John Hick, *Death and Eternal Life* (London: Collins, 1976).

36. Norman Vincent Peale, *The Power of Positive Thinking* (Surrey, U.K.: World's Work, 1953).

37. Gallup, *Adventures in Immortality*, p. 159.

38. David Royse, "The Near Death Experience: A Survey of Clergy's Attitudes and Knowledge," *The Journal of Pastoral Care* 39 (1985): 31–42.

39. Lori Bechtel, A. Chen, R. A. Pierce, and B. A. Walker, "Assessment of Clergy Knowledge and Attitudes Toward Near-Death Experiences," *Journal of Near-Death Studies* 10 (1992): 161–70.

40. Michael Perry, "Assessment of Clergy Knowledge and Attitudes" (letter), *Journal of Near-Death Studies* 11 (1992): 129.

41. Personal Correspondence Department, Worldwide Church of God, Form *Letter from Pastor General Joseph W. Tkach,* undated.

42. Gallup, *Adventures in Immortality*, p. 160.

43. Kerby Anderson, *Life, Death and Beyond* (Grand Rapids, Mich.: Zondervan, 1980), p. 122.

44. Ibid, p. 151.

45. Sabom, *Recollections of Death.*

46. Annalee Oakes, "Near Death Events and Critical Care Nursing," in Bruce Greyson and Charles Flynn (eds.), *The Near-Death Experience: Problems, Prospects and Perspectives* (Springfield, Ill.: Charles C Thomas, 1984), pp. 223–31.

47. Roberta Orne, "Nurse's Views of NDEs," *American Journal of Nursing* 86 (1986): 419–20.

48. Barbara Walker and Robert Russell, "Assessing Psychologists' Knowledge and Attitudes Toward Near-Death Phenomena," *Journal of Near-Death Studies* 8 (1989): 103–10.

49. Michael Grosso, "The Myth of the Near-Death Journey," *Journal of Near-Death Studies* 10 (1991): 49–60.

50. *Revitalized Signs: Newsletter for the International Association for Near-Death Studies.* Volumes 8 and 9 of this quarterly were perused for this information.

51. Robert Basil, "The Popular Appeal of the Near-Death Experience," *Journal of Near-Death Studies* 10 (1991): 61–68.

52. Ibid., p. 64.

CHAPTER 5

1. Readers Digest Association, *Folklore, Myths and Legends of Britain* (London: Readers Digest Association, 1977), p. 53.

2. Joseph Varon and George C. Sternbach, "Cardiopulmonary Resuscitation: Lessons from the Past," *Journal of Emergency Medicine* 9 (1991): 503.

3. Mary C. Vrtis, "Cost/Benefit Analysis of Cardiopulmonary Resuscitation: A History of CPR—Part 1," *Nursing Management* 23 (1992): 50.

4. Arlo S. Hermreck, "The History of Cardiopulmonary Resuscitation," *American Journal of Surgery* 156 (1988): 435.

5. Varon and Sternbach, "Cardiopulmonary Resuscitation."

6. John Paraskos, "Biblical Accounts of Resuscitation," *Journal of the History of Medicine and Allied Sciences,* 47 (1992): 310–21.

7. Hermreck, "Cardiopulmonary Resuscitation."

8. Peter Safar, "Initiation of Closed Chest Cardiopulmonary Resuscitation Basic Life Support: A Personal History," *Resuscitation* 18 (1989): 16.

9. George H. Buck, "Development of Simulators in Medical Education," *Gesnerus* 48 (1991): 15–16.

10. Ibid., pp 17–18.

11. Varon and Sternbach, "Cardiopulmonary Resuscitation," p. 507.

12. Robert M. Veatch, *Death, Dying and the Biological Revolution* (New Haven, Conn.: Yale University Press, 1976), p. 3.

13. Victor W. Marshall, *Last Chapters: A Sociology of Aging and Dying* (Monterey, Calif.: Brooks/Cole, 1980), pp. 18–24.

14. Joseph Eyer, "Prosperity as a Cause of Death," *International Journal of Health Services* 7 (1977): 125–50.

15. Ibid.

16. Hermreck, "Cardiopulmonary Resuscitation," p. 435.

17. Ibid.

18. Vrtis, "Cost/Benefit Analysis."

19. Ibid., p. 51.

20. For an extended discussion of this issue see Kellehear, "Are We a 'Death Denying' Society?"

21. Jessica Mitford, *The American Way of Death* (London: Quartet, 1980).

22. Robert Fulton and Greg Owen, "Death and Society in Twentieth Century America," *Omega* 18 (1988): 379–95.

23. David Armstrong, "Silence and Truth in Death and Dying," *Social Science and Medicine* 24 (1987): 651–57.

24. Fulton and Owen, "Death and Society," p. 381.

25. Michael Kearl, *Endings: A Sociology of Death and Dying* (New York: Oxford University Press, 1989), p. 383.

26. Erwin Ackerknecht, "Death in the History of Medicine," *Bulletin of the History of Medicine* 42 (1968): 19–23.

27. Phillippe Aries, *Western Attitudes Toward Death* (London: Johns Hopkins University Press, 1974).

28. Michael Simpson, *Dying, Death and Grief: A Critically Annotated Bibliography and Source Book of Thanatology and Terminal Care* (New York: Plenum Press, 1979).

29. Tony Walter, "Death in the New Age," *Religion* 23 (1993): 131.

30. Armstrong, "Silence and Truth."

31. Ibid., p. 655.

32. Elizabeth Kubler-Ross, *On Death and Dying* (New York: Macmillan, 1969).

33. Moody, *Life after Life.*

34. Sabom, *Recollections of Death.*

35. See Keith Roberts, *Religion in Sociological Perspective* (Homeward, Ill.: Dorsey Press, 1984), esp. pp. 373–404.

36. Ibid., p. 376.

37. Jerome R. Malino, "Coping with Death in Western Religious Civilization," *Zygon: Journal of Religion and Science* 1 (1966): 354–65.

38. Ibid., p. 354.

39. Ibid, p. 363.

40. Grey, *Return from Death.*

41. David Lorimer, *Whole in One: The Near-Death Experience and the Ethic of Interconnectedness* (London: Arkana, 1990).

42. Zaleski, *Otherworld Journeys.*

43. Christopher Lasch, *The Culture of Narcissism* (London: Abacus, 1980), p. xv.

44. John Naisbitt, *Megatrends* (London: MacDonald, 1984).

45. Ibid.

46. See Neil Smelser, *Theory of Collective Behavior* (New York: Free Press, 1963), a widely cited work still considered a classic. Also see Ron Roberts and Robert Kloss, *Social Movements* (London: C. V. Mosby, 1979).

47. Naisbitt, *Megatrends,* p. 42.

48. Lasch, *Culture of Narcissism.*

49. See Ivan Illich, *Limits to Medicine* (Harmondsworth, U.K.: Penguin, 1976); I. K. Zola, "Medicine as an Institution of Social Control," *Sociological Review* 20 (1972): 487–504; R. Taylor, *Medicine Out of Control* (Melbourne: Sun Books, 1979); and L. Doyal and J. Pennell, *The Political Economy of Health* (London: Pluto, 1979), to name only a few works still regularly cited today.

CHAPTER 6

1. See J. C. Davis, *Utopia and the Ideal Society* (Cambridge: Cambridge University Press, 1981); J. C. Davis, "The History of Utopia: The Chronology of No-

where," in P. Alexander and R. Gill (eds.), *Utopias* (London: Duckworth, 1984), pp 1–17.

2. Greyson, "A Typology of Near-Death Experiences."

3. Sabom, *Recollections of Death.*

4. J. Lindley, S. Bryan, and B. Conley, "Near-Death Experiences in a Pacific Northwest American Population: The Evergreen Study," *Anabiosis* 1 (1981): 104–24.

5. Ring, *Life at Death.*

6. Gallup, *Adventures in Immortality.*

7. Moody, *Life after Life.*

8. Ring, *Life at Death.*

9. Sabom, *Recollections of Death.*

10. Grey, *Return from Death.*

11. Zaleski, *Otherworld Journeys.*

12. Lundahl, "Perceived Other World."

13. Grey, *Return from Death,* p. 81.

14. Ibid., p. 49.

15. Eadie, *Embraced,* p. 78.

16. Ibid., p. 76.

17. Elder, *And When I Die, Will I Be Dead?* p. 25.

18. Ritchie. *Return from Tomorrow,* p. 70.

19. Ibid., p. 72.

20. Ibid., p. 69.

21. Eadie, *Embraced,* pp. 74–75.

22. Lundahl, "Perceived Other World."

23. Ibid.

24. Counts, "Near-Death and Out-of-Body Experiences."

25. Ring, *Heading Toward Omega,* p. 91.

26. F. E. Manuel and F. R. Manuel, *Utopian Thought in the Western World* (Oxford: Basil Blackwell, 1979).

27. P. Alexander and R. Gill (eds.), *Utopias* (London: Duckworth, 1984).

28. K. Kumar, *Utopia and Anti-utopia in Modern Times* (Oxford: Basil Blackwell, 1987).

29. Davis, *Utopia and the Ideal Society.*

30. K. Mannheim, *Ideology and Utopia* (London: Routledge and Kegan Paul, 1960).

31. K. S. Walters, *The Sane Society Ideal in Modern Utopianism* (Toronto: Edwin Mellen Press, 1989).

32. Alexander, "Grimm's Utopia."

33. M. Weber, *The Theory of Social and Economic Organization* (New York: Free Press, 1947).

34. Davis, *Utopia and the Ideal Society.*

35. Ibid., p. 21.

36. P. Brookesmith, *Life after Death* (London: Orbis, 1984).

37. Grey, *Return from Death,* p. 54.

38. Davis, *Utopia and the Ideal Society*.

39. Ibid.

40. Ibid.

41. Ibid.

42. Counts, "Near-Death and Out-of-Body Experiences."

43. Lundahl, "Perceived Other World," pp. 323–24.

44. Davis, *The History of Utopia*.

45. Davis, *Utopia and the Ideal Society*.

46. Ring, *Heading Toward Omega;* and K. Ring, "Prophetic Visions in 1988: A Critical Reappraisal," *Journal of Near-Death Studies* 7 (1988): 4–18.

47. J. F. C. Harrison, "Millennium and Utopia," in Alexander and Gill (eds.), *Utopias*.

48. Ibid.

49. Davis, *Utopia and the Ideal Society*, p. 38.

50. Ibid, p. 388.

51. Davis, *The History of Utopia*.

52. B. Goodwin, "Economic and Social Innovation in Utopia," in Alexander and Gill (eds.), *Utopias*.

53. Alexander, "Grimm's Utopia."

54. E. Bloch, *The Utopian Function of Art and Literature: Selected Essays* (Cambridge, Mass.: MIT Press, 1988).

55. Kumar, *Utopia and Anti-Utopia*.

56. Alexander, "Grimm's Utopia."

57. M. Bradbury, "Postmodernism," in A. Bullock, S. Trombley, and B. Eadie (eds.), *The Harper Dictionary of Modern Thought* (New York: Harper and Row, 1988), pp. 671–72.

58. Zaleski, *Otherworld Journeys*.

59. Ibid.; and E. W. Fenske, "The Near-Death Experience: An Ancient Truth, a Modern Mystery," *Journal of Near-Death Studies* 8 (1990): 129–48.

60. Kumar, *Utopia and Anti-Utopia*.

61. Ibid.

62. Walters, *Sane Society Ideal*.

63. Zaleski, *Otherworld Journeys*.

64. Counts, "Near-Death and Out-of-Body Experiences."

65. Alexander and Gill, *Utopias*.

66. Davis, *Utopia and the Ideal Society*, p. 320.

67. Alexander and Gill, *Utopias*.

68. P. Beilharz, "Utopia and Its Futures," *Thesis Eleven* 24 (1989): 150–60.

CHAPTER 7

1. C. Zaleski, *Otherworld Journeys*, p. 165.

2. See, for example, G. O. Gabbard, S. W. Twemlow, and F. C. Jones. "Do

'Near-Death Experiences' Occur Only Near Death?" *Journal of Nervous and Mental Disease* 169 (1981): 374–77; B. C. Bates and A. Stanley, "The Epidemiology and Differential Diagnosis of Near-Death Experience," *American Journal of Orthopsychiatry* 55 (1985): 542–49; Stevenson, Cook, and McClean-Rice, "Are Persons Reporting 'Near-Death Experiences' Really Near Death?"

3. D. B. Carr, "Endorphins at the Approach of Death," *Lancet* (14 Feb. 1981): 390. Carr does not offer much evidence for this belief, merely linking Osis and Haraldsson's (1977) description of NDEs and deathbed visions observed by health professionals to his assertion that these visions are endorphin related. Osis and Haraldsson themselves do not make the connection with endorphins.

4. Blackmore, *Dying to Live,* pp. 106–110.

5. Seigal, "Psychology of Life after Death,"

6. H. Kluver, *Mescal and Mechanisms of Hallucination* (Chicago: University of Chicago Press, 1967). Note, however, that Zaleski (1987: 243) makes the methodologically interesting point that Seigal trained his subjects to use these descriptive codes. Those subjects who were untrained in this convention were less "successful" in reducing their impression. Thus it appears that Seigal's results in this area are not derived inductively, a technical problem that must cause us to question the reliability and validity of the conclusions. This, however, may not be a problem confined to Seigal. Seigal, and later Blackmore (*Dying to Live*) and J. D. Cowan ("Spontaneous Symmetry Breaking in Large Scale Nervous Activity," *International Journal of Quantum Chemistry* 22 [1982]: 1059–82), all refer to the early work of Kluver (1967). But Kluver himself does not describe his methodology in detail. His "form constants" appear to be merely his impressions from various published accounts and personal experiences with Mescalan. Even here, he notes that there are other ways of classifying constants and cites Rouhier's attempt as an example (A. Rouhier, *La Plante qui fait les yeux emerveilles; le peyoti* [*Echinocactus Williamsii Lem*] [Paris: Doin, 1927]).

7. V. P. Neppe, "Near-Death Experiences: A New Challenge in Temporal Lobe Phenomenology? Comments on 'A Neurobiological Model for the Near-Death Experience.'" *Journal of Near-Death Studies* 7 (1989): 243–47.

8. Carr, "Endorphins."

9. J. P. Jourdan, "Near-Death and Transcendental Experiences: Neurophysiological Correlates of Mystical Traditions," *Journal of Near-Death Studies* 12 (1994): 198.

10. Zaleski, *Otherworld Journeys,* p. 180.

11. I. N. Bulhof, *The Language of Science: A Study of the Relationship Between Literature and Science in the Perspective of a Hermeneutical Ontology* (Leiden: E. J. Brill), p. 58.

12. C. Wright Mills, "Language, Logic and Culture," in I. L. Horowitz (ed.), *Power, Politics and People: The Collected Essays of C. Wright Mills* (New York: Oxford University Press, 1963), p. 433.

13. Seigal, "The Psychology of Life after Death," p. 919.

14. Ibid., p. 914. Emphasis added.

15. Ibid., p. 916.

16. Ibid., p. 922.

17. Ibid., p. 921.

18. Ibid.

19. S. J. Blackmore, "Glimpse of an Afterlife—or Just the Dying Brain?" *Psi Researcher* 6 (1992): 3. Emphasis added.

20. K. Popper, *The Logic of Scientific Discovery* (London: Hutchinson, 1968).

21. T. S. Kuhn, *The Copernican Revolution* (New York: Random House, 1959).

22. M. Foucault, *The Archaeology of Knowledge* (London: Tavistock, 1974).

23. L. Fleck, *Genesis and Development of a Scientific Fact* (Chicago: University of Chicago Press, 1979).

24. Saavedra Aguila and Gomez-Jeria, "Neurobiological Model."

25. Ibid., p. 205.

26. Ibid., p. 209.

27. Ibid., p. 212.

28. Ibid., p. 214. Emphasis added.

29. R. Noyes, "Comments on 'A Neurobiological Model for Near-Death Experiences,' " *Journal of Near-Death Studies* 7 (1989): 249.

30. Saavedra-Aguila and Gomez-Jeria, "Neurobiological Model," p. 266.

31. S. J. Blackmore, "Blackmore's Reply to Fontana," *Psi Researcher* 7 (1992): 7.

32. A. F. Chalmers, *What Is This Thing Called Science?* (St Lucia: University of Queensland Press, 1976).

33. Bulhof, *Language of Science,* p. 11.

34. J. Ellard, "Schizophrenia: Here Today, Gone Tomorrow," *Modern Medicine of Australia* 28 (Jan. 1985): 9–13.

35. M. Nicholson and C. McLaughlin, "Social Constructionism and Medical Sociology: A Study of the Vascular Theory of Multiple Sclerosis," *Sociology of Health and Illness* 10 (1988): 234–61.

36. M. A. Persinger, "Modern Neuroscience and Near-Death Experiences: Expectancies and Implications. Comments on 'A Neurobiological Model for Near-Death Experiences.' " *Journal of Near-Death Studies* 7 (1989): 237.

37. Ibid.

38. D. M. Bear and P. Fedio, "Quantitative Analysis of Interictal Behavior in Temporal Lobe Epilepsy," *Archives of Neurology* 34 (1977): 464.

39. Saavedra-Aguila and Gomez-Jeria, "Neurobiological Model," p. 218.

40. J. Kroll and B. Bachrach, "Visions and Psychopathology in the Middle Ages," *Journal of Nervous and Mental Disease* 170 (1982): 41–49.

41. D. Landsborough, "St. Paul and Temporal Lobe Epilepsy," *Journal of Neurology, Neurosurgery and Psychiatry* 50 (1987): 659.

42. D. Janz, "Epilepsy, Viewed Metaphysically: An Interpretation of the Biblical Story of the Epileptic Boy and of Raphael's Transfiguration," *Epilepsia* 27 (1986): 316–22.

43. Kroll and Bachrach, "Visions and Psychopathology," p. 47.

44. G. Roberts and J. Owen, "The Near-Death Experience," *British Journal of Psychiatry* 153 (1988): 607–17.

45. Kung, *Eternal Life?*

46. Zaleski, *Otherworld Journeys,* p. 164.

47. Seigal, "The Psychology of Life after Death," p. 922.

48. W. F. Matchett, "Repeated Hallucinatory Experiences as a Part of the Mourning Process among the Hopi Indian Women," *Psychiatry* 35 (1972): 185–94.

49. N. D. Jewson, "The Disappearance of the Sick Man from Medical Cosmology, 1770–1870," *Sociology* 10 (1976): 225–44.

50. Ibid., p. 226.

51. M. Morse, "Comments on 'A Neurobiological Model for Near-Death Experiences,' " *Journal of Near-Death Studies* 7 (1989): 224.

52. D. C. Dennett, *Consciousness Explained* (Boston: Little, Brown, 1991), p. 455.

53. Seigal, "The Psychology of Life after Death," pp. 926–27.

54. Ibid. Emphasis added.

55. Ibid., p. 928.

56. Blackmore and Troscianko, "Physiology of the Tunnel," p. 26. Emphasis added.

57. Ibid., p. 23.

58. B. S. Turner, *Regulating Bodies: Essays in Medical Sociology* (London: Routledge, 1992), p. 142.

59. Hansen, "CSICOP and Skepticism."

60. Blackmore and Troscianko, "Physiology of the Tunnel."

61. Roberts and Owen, "The Near-Death Experience."

62. Seigal,. "The Psychology of Life after Death."

63. Saavedra-Aguila and Gomez-Jeria, "Neurobiological Model."

64. D. E. Leary, "Communication, Persuasion and the Establishment of Academic Disciplines: The Case of American Psychology," in R. H. Brown (ed.), *Writing the Social Text: Poetics and Politics in Social Science Discourse* (New York: Aldine, 1992), pp. 73–90.

65. Jewson, "Disappearance."

66. I. K. Zola, "Medicine as an Institution of Social Control," *Sociological Review* 20 (1972): 487–504.

67. Leary,. "Communication."

68. Zola, "Medicine," p. 496.

69. M. Polanyi, *Personal Knowledge* (London: Routledge and Kegan Paul, 1973).

70. There is a vast literature on this topic. Representatives of this general argument are E. Friedson, *Profession of Medicine* (New York: Dodd, Mead, 1970); P. Starr, *The Social Transformation of American Medicine* (New York: Basic Books, 1982); I. Illich, *Limits to Medicine* (Harmondsworth, U.K.: Penguin, 1976); and E. Willis, *Medical Dominance* (Sydney: Allen & Unwin, 1989).

71. K. White and E. Willis, "The Languages of AIDS," *New Zealand Sociology* 7 (1992): 127–49.

72. J. Powles, "On the Limitations of Modern Medicine," *Science, Medicine and Man* 1 (1973): 1–30.

73. White and Willis. "Languages of AIDS," p. 128.

74. Walter, "Death in the New Age."

75. Jewson, "Disappearance."

76. Ibid., pp. 233–34.

77. Ibid., p. 228.

78. Kroll and Bachrach, "Visions and Psychopathology," p. 47.

79. Turner, *Regulating Bodies*, p. 142.

80. Nicholson and McLaughlin, "Social Constructionism," p. 250.

CHAPTER 8

1. M. Bloch and J. Parry (eds.), *Death and the Regeneration of Life* (Cambridge: Cambridge University Press, 1982).

2. D. Lorimer, *Whole in One: The Near-Death Experience and the Ethic of Interconnectedness* (London: Arkana, 1990).

3. See, for example, B. Bettleheim, *The Uses of Enchantment: The Meaning and Importance of Fairy Tales* (Harmondsworth, U.K.: Penguin, 1978); and M. Rustin and M. Rustin, *Narratives of Loss and Love: Studies in Modern Children's Fiction* (London: Verso, 1987).

4. Williams, *The Velveteen Rabbit*.

5. S. V. Daniels, "The Velveteen Rabbit: A Kleinian Perspective," *Children's Literature: An International Journal* 18 (1990): 17–30.

6. J. C. Stott, *Children's Literature from A to Z: A Guide for Parents and Teachers* (New York: McGraw-Hill, 1984).

7. G. Weales, "Childrens Books," *New York Times Book Review* (3 Ap. 1983): 3.

8. M. Fisher, "Review of 'The Velveteen Rabbit.'" *Growing Point* 9 (1971): 1657.

9. Williams, *The Velveteen Rabbit*, 16–17.

10. Ibid., p. 24.

11. Ibid., p. 30.

12. Ibid., p. 34.

13. Ibid., p. 37.

14. Ibid., p. 40.

15. F. McNulty, "Children's Books for Christmas," *New Yorker* (6 Dec. 1992): 176–82.

16. Weales, "Children's Books."

17. McNulty. "Children's Books for Christmas," p. 180.

18. Daniels, "The Velveteen Rabbit," p. 28.

19. Ibid., pp. 17–18.

20. Ibid., p. 27.

21. Ibid., p. 27.

22. A. Game, *Undoing the Social: Toward a Deconstructive Sociology* (Buckingham, U.K.: Open University Press, 1991), p. 5.

23. Williams, *The Velveteen Rabbit,* pp. 28–29.

24. R. M. Lockley, *The Private Life of the Rabbit: An Account of the Life History and Social Behaviour of the Wild Rabbit* (London: Andre Deutsch, 1976), p. 142.

25. S. Freud, *Beyond the Pleasure Principle* (London: Hogath Press, 1961).

26. Tolstoy, *Death of Ivan Ilyich.*

27. Williams, The Velveteen Rabbit, 35–36.

28. Ibid., p. 37.

29. Ibid., p. 38.

30. Ibid., p. 28.

31. Weales, "Children's Books."

32. H. Ferris, *Writing Books for Boys and Girls* (New York: Doubleday, 1952), p. 21.

33. Williams, *The Velveteen Rabbit,* p. 38.

34. Ibid., p. 17.

35. Zaleski, *Otherworld Journeys,* p. 193.

36. Quoted in Stott, *Children's Literature.*

37. Williams, *The Velveteen Rabbit,* p. 44.

38. See, for example, C. P. Flynn, "Death and the Primary of Love in Works of Dickens, Hugo and Wilder," *Anabiosis* 4 (1984): 125–41; S. Straight, "A Wave Among Wave: Katherine Anne Porter's Near-Death Experience," *Anabiosis* 4 (1984): 107–23; J. Wren-Lewis, "Joy without a Cause: An Anticipation of Modern 'Near-Death Experiences' Research in G. K. Chesterton's Novel *The Ball and the Cross.*" *The Chesterton Review* 12 (1986): 49–61.

39. Ring, "Amazing Grace."

40. Zaleski, *Otherworld Journeys.*

41. J. F. Lyotard, *The Postmodern Condition: A Report on Knowledge* (Manchester, U.K.: Manchester University Press, 1984), p. 81.

42. Morse, "Comments."

43. F. Baum, *The Wizard of Oz* (London: Chancellor, 1987).

44. Carrol, *Alice's Adventures in Wonderland.*

45. F. G. Greene, "Motifs of Passage into Worlds Imaginary and Fantastic," *Journal of Near-Death Studies* 10 (1992): 205–31.

46. Zaleski, *Otherworld Journeys.*

47. Ibid., p. 191.

CHAPTER 9

1. J. K. W. Morrice, *Crisis Intervention—Studies in Community Care* (Oxford: Pergamon, 1976), p. 20.

2. B. Raphael, *When Disaster Strikes* (New York: Basic Books, 1986), p. 5.

3. Butler, "Life Review," p. 73.

4. For a good general review of the area, see Suedfeld and Mocellin, "The Sensed Presence."

5. Ring, *Life at Death*, p. 40.

6. R. Moody and P. Perry, *Reunions: Visionary Encounters with Departed Loved Ones* (New York: Villard, 1993), p. x. Also see R. A. Kalish and D. K. Reynolds, "Phenomenological Reality and Postdeath Contact," *Journal for the Scientific Study of Religion* 12 (1973): 209–21; and A. M. Greeley, "Hallucinations Among the Widowed," *Sociology and Social Research* 74 (1987): 258–65.

7. Moody, *Life after Life*, p. 23.

8. Kearl, *Endings*, pp. 493–96.

9. Ibid., p. 496.

10. Mary Midgley, *Science as Salvation: A Modern Myth and Its Meaning* (London: Routledge, 1992), pp. 17–21.

11. Ibid., p. 58.

12. Stephen Toulmin, *Cosmopolis: The Hidden Agenda of Modernity* (New York: Free Press, 1990).

13. Ibid., p. 30.

14. Ibid., pp. 28–42.

15. Ibid., p. 185.

16. Kubler-Ross, *On Death and Dying*.

17. Midgley, *Science as Salvation*, p. 51.

18. Zaleski, *Otherworld Journeys*, p. 180.

19. Mills, *The Sociological Imagination*.

APPENDIX

1. For a fine discussion of how the experience of travel is mediated by a changing sense of self see, M. Neumann's essay "The Trial Through Experience: Finding Self in the Recollection of Travel," in C. Ellis and M. G. Flaherty (eds.), *Investigating Subjectivity: Research on Lived Experiences.* (Newbury Park, Calif.: Sage, 1992), pp. 176–201. This essay also has some important implications for understanding the role of recounting NDEs in the social construction of self.

2. Kai Erikson, *Wayward Puritans: A Study in the Sociology of Deviance* (New York: Wiley, 1966), p. 6.

3. Eadie, *Embraced by the Light*, p. 14.

4. See Patricia Weibust, "Being One with God Is Something That Can Be Done without Rules: Commentary on Allan Kellehear's 'Near-Death Experiences and the Pursuit of the Ideal Society,' " *Journal of Near-Death Studies* 10 (1991): 107–11.

5. Elder, *And When I Die, Will I Be Dead?*

6. Ring, *Life at Death*, pp. 68–69, and his more recent essay, "Amazing Grace."

7. Ritchie, *Return from Tomorrow*.

8. R. Neihardt, *Black Elk Speaks* (Lincoln: University of Nebraska Press, 1961).

9. Eadie, *Embraced by the Light.*

10. See Antonia Mills, "Commentary on Allan Kellehear's 'Near-Death Experiences and the Pursuit of the Ideal Society,' " *Journal of Near-Death Studies* 10 (1991): 113–22.

11. Ritchie, *Return from Tomorrow,* p. 69.

12. Ibid., p. 70.

13. Ibid., p. 72.

14. Eadie, *Embraced by the Light,* pp. 54–61.

15. Ibid., p. 55.

16. Ibid., p. 75.

17. Moody, *Light after Life,* p. 57.

18. Ibid., p. 56.

19. Ibid., p. 51.

20. Allan Kellehear, *The Unobtrusive Researcher: A Guide to Methods* (Sydney: Allen & Unwin, 1993), p. 1.

21. Greyson and Bush, "Distressing Near-Death Experiences."

22. Zaleski, *Otherworld Journeys.*

23. Lorimer, *Whole in One.*

24. Blackmore, *Dying to Live.*

25. Blackmore, "Near Death Experiences in India."

26. Osis and Haraldsson, *At the Hour of Death.*

27. Pasricha and Stevenson, "Near-Death Experiences in India."

28. Pasricha, "A Systematic Survey of Near-Death Experiences in South India."

29. Kellehear, Stevenson, Pasricha, and Cook, "Tunnel Sensation."

30. Zaleski, *Otherworld Journeys.*

BIBLIOGRAPHY

Ackerknecht, E. (1968). "Death in the History of Medicine." *Bulletin of the History of Medicine* 42: 19–23.

Alexander, P. (1984). "Grimm's Utopia: Motives and Justifications." In P. Alexander and R. Gill (eds.), *Utopias* (pp. 31–42). London: Duckworth.

Alexander, P., and Gill, R. (eds.). (1984). *Utopias.* London: Duckworth.

Allport, G. (1958). *The Nature of Prejudice.* New York: Doubleday.

Anderson, K. (1980). *Life, Death and Beyond.* Grand Rapids, Mich.: Zondervan.

Aries, P. (1974). *Western Attitudes Toward Death.* London: Johns Hopkins University Press.

Armstrong, D. (1987). "Silence and Truth in Death and Dying." *Social Science and Medicine* 24: 651–57.

Atwater, P. M. H. (1988). *Coming Back to Life: The After-Effects of the Near-Death Experience.* New York: Ballantine.

Bailey, M., and Bailey, M. (1974). *Staying Alive!* New York: David McKay.

Basil, R. (ed.). (1989). *Not Necessarily the New Age: Critical Essays.* New York: Prometheus.

Basil, R. (1991). "The Popular Appeal of the Near-Death Experience." *Journal of Near-Death Studies* 10: 61–68.

Bates, B. C., and Stanley, A. (1985). "The Epidemiology and Differential Diagnosis of Near-Death Experience." *American Journal of Orthopsychiatry* 55: 542–49.

Baum, F. (1987). *The Wizard of Oz.* London: Chancellor.

Bear, D. M., and Fedio, P. (1977). "Quantitative Analysis of Interictal Behavior in Temporal Lobe Epilepsy." *Archives of Neurology* 34: 454–67.

Bechtel, L., Chen, A., Pierce, R. A., and Walker, B. A. (1992). "Assessment of Clergy Knowledge and Attitudes Toward Near-Death Experiences." *Journal of Near-Death Studies* 10: 161–170.

Becker, C. B. (1981). "The Centrality of Near-Death Experiences in Chinese Pure Land Buddhism." *Anabiosis* 1: 154–71.

Becker, C. B. (1982). "The Failure of Saganomics: Why Birth Models Cannot Explain Near-Death Phenomena." *Anabiosis* 2: 102–9.

Becker, C. B. (1984). "The Pure Land Re-visited: Sino-Japanese Meditations and Near-Death Experiences of the Next World." *Anabiosis* 4: 51–68.

Becker, E. (1973). *The Denial of Death*. New York: Collier-Macmillan.

Beilharz, P. (1989). "Utopia and Its Futures." *Thesis Eleven* 24: 150–60.

Bellah, R. N. (1976). *Beyond Belief: Essays on Religion in a Post-Traditional World*. New York: Harper and Row.

Berndt, R. M., and Berndt, C. H. (1989). *The Speaking Land: Myth and Story in Aboriginal Australia*. Harmondsworth, U.K.: Penguin.

Bettelheim. B. (1978). *The Uses of Enchantment: The Meaning and Importance of Fairy Tales*. Harmondsworth, U.K.: Penguin.

Blackmore, S. J. (1984). "Birth and the OBE: An Unhelpful Analogy." *Journal of the American Society for Psychical Research* 52: 225–44.

Blackmore, S. J. (1992). "Blackmore's Reply to Fontana." *Psi Researcher* 7: 6–7.

Blackmore, S. J. (1992). "Glimpse of an Afterlife—or Just the Dying Brain?" *Psi Researcher* 6: 2–3.

Blackmore, S. J. (1993). *Dying to Live: Science and the Near-Death Experience*. London: Grafton.

Blackmore, S. J. (1993). "Near-Death Experiences in India: They Have Tunnels Too." *Journal of Near-Death Studies* 11: 205–17.

Blackmore, S. J., and Troscianko, T. S. (1989). "The Physiology of the Tunnel." *Journal of Near-Death Studies* 8: 15–28.

Bloch, E. (1988). *The Utopian Function of Art and Literature: Selected Essays*. Cambridge, Mass.: MIT Press.

Bloch, M., and Parry, J. (Eds) (1982). *Death and the Regeneration of Life*. Cambridge: Cambridge University Press.

Bradbury, M. (1988). "Post-modernism." In A. Bullock, S. Trombley, and B. Eadie (eds.), *The Harper Dictionary of Modern Thought* (pp. 671–72). New York: Harper and Row.

Brookesmith, P. (1984). *Life After Death*. London: Orbis.

Brown, N. O. (1968). *Life Against Death: The Psychoanalytic Meaning of History*. London: Sphere.

Buck, G. (1991). "Development of Simulators in Medical Education." *Gesnerus* 48: 7–28.

Bulhof, I. N. (1992). *The Language of Science: A Study of the Relationship Between Literature and Science in the Perspective of a Hermeneutical Ontology*. Leiden: E. J. Brill.

Bush, N. E. (1983). "The Near-Death Experience in Children: Shades of the Prison House Re-Opening." *Anabiosis* 3: 177–93.

Butler, R. N. (1963). "The Life Review: An Integration of Reminiscence in the Aged." *Psychiatry* 26: 65–76.

Carr, D. B. (1981, 14. Feb.). "Endorphins at the Approach of Death." *Lancet:* 390.

Carroll, L. (1965). *Alice's Adventure in Wonderland*. New York: Airmont.

Chalmers, A. F. (1976). *What Is This Thing Called Science?* St. Lucia: University of Queensland Press.

Counts, D. A. (1983). "Near-Death and Out-of-Body Experiences in a Melanesian Society." *Anabiosis* 3: 115–35.

Cowan, J. D. (1982). "Spontaneous Symmetry Breaking in Large Scale Nervous Activity." *International Journal of Quantum Chemistry* 22: 1059–82.

Culver, J. (1989, 30 Sept.–1 Oct.). "Ordeal at Sea." *The Australian Magazine:* 8–14.

Daniels, S. V. (1990). "The Velveteen Rabbit: A Kleinian Perspective." *Children's Literature: An International Journal* 18: 17–30.

Davis, J. C. (1981). *Utopia and the Ideal Society.* Cambridge: Cambridge University Press.

Davis, J. C. (1984). "The History of Utopia: The Chronology of Nowhere." In P. Alexander and R. Gill (eds.), *Utopias* (pp. 1–17). London: Duckworth.

Dennett, D. C. (1991). *Consciousness Explained.* Boston: Little, Brown.

Douglas, M. (1966). *Purity and Danger.* London: Routledge and Kegan Paul.

Doyal, L., and Pennell, J. (1979). *The Political Economy of Health.* London: Pluto.

Drab, K. J. (1981). "Unresolved Problems in the Study of Near-Death Experiences: Some Suggestions for Research and Theory." *Anabiosis* 1: 27–43.

Drab, K. J. (1981). "The Tunnel Experience: Reality or Experience?" *Anabiosis* 1: 126–52.

Durkheim, E. (1965). *The Elementary Forms of the Religious Life.* New York: Free Press.

Eadie, B. (1992). *Embraced by the Light.* Placerville, Calif.: Gold Leaf Press.

Elder, B. (1987). *And When I Die, Will I Be Dead?* Sydney: Australian Broadcasting Commission.

Ellard, J. (1985, January). "Schizophrenia: Here Today, Gone Tomorrow." *Modern Medicine of Australia* 28: 9–13.

Erikson, K. (1966). *Wayward Puritans: A Study in the Sociology of Deviance.* New York: Wiley.

Evans-Wentz, W. Y. (1960). *The Tibetan Book of the Dead.* London: Oxford University Press.

Eyer, J. (1977). "Prosperity as a Cause of Death." *International Journal of Health Services* 7: 125–50.

Fenske, E. W. (1990). "The Near-Death Experience: An Ancient Truth, a Modern Mystery." *Journal of Near-Death Sudies* 8: 129–48.

Ferris, H. (1952). *Writing Books for Boys and Girls.* New York: Doubleday.

Fisher, M. (1971). Review of "The Velveteen Rabbit." *Growing Point* 9: 1657.

Fleck, L. (1979). (orig. 1935). *Genesis and Development of a Scientific Fact.* Chicago: University of Chicago Press.

Flynn, C. P. (1984). "Death and the Primacy of Love in Works of Dickens, Hugo, and Wilder." *Anabiosis* 4: 125–41.

Flynn, C. P. (1986). *After the Beyond: Human Transformation and the Near-Death Experience.* Englewood Cliffs, N.J.: Prentice-Hall.

Foucault, M. (1974). *The Archaeology of Knowledge.* London: Tavistock.

Freud, S. (1961). *Beyond the Pleasure Principle.* London: Hogarth Press.

Friedson, E. (1970). *Profession of Medicine.* New York: Dodd, Mead.

Fulton, R., and Owen, G. (1988). "Death and Society in Twentieth Century America." *Omega* 18: 379–95.

Gabbard, G. O., Twemlow, S. W., and Jones, F. C. (1981). "Do 'near-death experiences' Occur Only Near Death?" *Journal of Nervous and Mental Disease* 169: 374–77.

Gallup, G., Jr., and Proctor, W. (1982). *Adventures in Immortality: A Look Beyond the Threshold of Death.* New York: McGraw-Hill.

Game, A. (1991). *Undoing the Social: Toward a Deconstructive Sociology.* Buckingham, U.K.: Open University Press.

Glaser, B. G., and Strauss, A. L. (1965). *Awareness of Dying.* New York: Aldine.

Glaser, B. G., and Strauss, A. L. (1968). *Time for Dying.* Chicago: Aldine.

Glaser, B. G., and Strauss, A. L. (1971). *Status Passage.* London: Routledge and Kegan Paul.

Gomez-Jeria, Juan S. (1993). "A Near-Death Experience Among the Mapuche People." *Journal of Near-Death Studies* 11: 219–22.

Goodwin, B. (1984). "Economic and Social Innovation in Utopia." In P. Alexander and R. Gill (eds.), *Utopias* (pp. 69–83). London: Duckworth.

Greeley, A. M. (1987). "Hallucinations Among the Widowed." *Sociology and Social Research* 74: 258–65.

Green, J. T. (1984). "Near-Death Experiences in a Chommorro Culture." *Vital Signs* 4: 1–2, 6–7.

Greene, F. G. (1992). "Motifs of Passage into Worlds Imaginary and Fantastic." *Journal of Near-Death Studies* 10: 205–31.

Grey, M. (1985). *Return from Death: An Exploration of the Near-Death Experience.* London: Arkana.

Greyson, B. (1981). "Near-Death Experiencers and Attempted Suicide." *Suicide and Life Threatening Behavior* 2: 10–16.

Greyson, B. (1981). "Toward a Psychological Explanation of Near Death Experiences: A Response to Dr. Grosso's Paper." *Anabiosis* 1: 88–103.

Greyson, B. (1988). "A Typology of Near-Death Experiences." *American Journal of Psychiatry* 142: 967–69.

Greyson, B., and Evans-Bush, N. (1992). "Distressing Near-Death Experiences." *Psychiatry* 55: 95–110.

Greyson, B., and Harris, B. (1987). "Clinical Approaches to the Near-Death Experiencer." *Journal of Near-Death Studies* 6: 41–52.

Greyson, B., and Stevenson, I. (1980). "The Phenomenology of Near-Death Experiences." *American Journal of Psychiatry* 137: 1193–96.

Grosso, M. (1981). "Toward an Explanation of Near-Death Phenomena." *Journal of the American Society for Psychical Research* 75: 37–60.

Grosso, M. (1991). "The Myth of the Near-Death Journey." *Journal of Near-Death Studies* 10: 49–60.

Hansen, G. P. (1992). "CSICOP and Skepticism: An Emerging Social Movement." *Journal of the American Society for Psychical Research* 86: 19–63.

Harrison, J. F. C. (1984). "Millennium and Utopia." In P. Alexander and R. Gill (eds.), *Utopias* (pp. 61–66). London: Duckworth.

Hearney, J. J. (1984). *The Sacred and the Psychic.* New York: Paulist Press.

Hermann, E. J. (1990). "The Near-Death Experience and the Taoism of Chuang Tzu." *Journal of Near-Death Studies* 8: 175–90.

Hermreck, A. (1988). "The History of Cardiopulmonary Resuscitation." *American Journal of Surgery* 156: 430–36.

Hick, J. (1976). *Death and Eternal Life*. London: Collins.

Illich, I. (1976). *Limits to Medicine*. Harmondsworth, U.K.: Penguin.

Insinger, M. (1991). "The Impact of Near-Death Experience on Family Relationships." *Journal of Near-Death Studies* 9: 141–81.

Janz, D. (1986). "Epilepsy, Viewed Metaphysically: An Interpretation of the Biblical Story of the Epileptic Boy and of Raphael's Transfiguration." *Epilepsia* 27: 316–22.

Jewson, N. D. (1976). "The Disappearance of the Sick Man from Medical Cosmology, 1770–1870." *Sociology* 10: 225–44.

Jourdan, J. P. (1994). "Near-Death and Transcendental Experiences: Neurophysiological Correlates of Mystical Traditions." *Journal of Near-Death Studies* 12: 177–200.

Kalish, R. A. (1970). "Non-Medical Interventions in Life and Death." *Social Science and Medicine* 4: 655–65.

Kalish, R. A., and Reynolds, D. K. (1973). "Phenomenological Reality and Postdeath Contact." *Journal for the Scientific Study of Religion* 12: 209–21.

Kastenbaum, R. (1984). *Is There Life after Death?* London: Prentice-Hall.

Kearl, M. (1989). *Endings: A Sociology of Death and Dying*. New York: Oxford University Press.

Kellehear, A. (1984). "Are We a Death-Denying Society? A Sociological Review." *Social Science and Medicine* 18: 713–23.

Kellehear, A. (1990). *Dying of Cancer: The Final Year of Life*. London: Harwood.

Kellehear, A. (1990). "The Near-Death Experience as Status Passage." *Social Science and Medicine* 31: 933–39.

Kellehear, A. (1993). *The Unobtrusive Researcher: A Guide to Methods*. Sydney: Allen & Unwin.

Kellehear, A., and Heaven, P. (1989). "Community Attitudes Toward the Near-Death Experience: An Australian Study." *Journal of Near-Death Studies* 7: 165–72.

Kellehear, A., Heaven, P., and Gao, J. (1990). "Community Attitudes Toward Near-Death Experiences: A Chinese Study." *Journal of Near-Death Studies* 8: 163–73.

Kellehear, A., Stevenson, I., Pasricha, S., and Cook, E. (1994). "The Absence of Tunnel Sensation in Near-Death Experiences from India." *Journal of Near-Death Studies* 13: 109–13.

King, M. (1985). *Being Pakeha: An Encounter with New Zealand and the Maori Renaissance*. Auckland, N.Z.: Hodder & Stoughton.

Kluver, H. (1967). *Mescal and Mechanisms of Hallucination*. Chicago: University of Chicago Press.

Kroll, J., and Bachrach, B. (1982). "Visions and Psychopathology in the Middle Ages." *Journal of Nervous and Mental Disease* 170: 41–49.

Kubler-Ross, E. (1969). *On Death and Dying.* New York: Macmillan.

Kuhn, T. S. (1959). *The Copernican Revolution.* New York: Random House.

Kumar, K. (1987). *Utopia and Anti-Utopia in Modern Times.* Oxford: Basil Black-well.

Kung, H. (1984). *Eternal Life?* London: Collins.

Lai, W. (in press). "Tales of Rebirths and the Later Pure Land Tradition in China." In M. Solomon (ed.), *Berkeley Buddhist Studies* Series 3. Berkeley, Calif.: Asian Humanities Press.

Landsborough, D. (1987). "St. Paul and Temporal Lobe Epilepsy." *Journal of Neurology, Neurosurgery and Psychiatry* 50: 659–64.

Lasch, C. (1980). *The Culture of Narcissism.* London: Abacus.

Leary, D. E. (1992). "Communication, Persuasion, and the Establishment of Academic Disciplines: The Case of American Psychology." In R. H. Brown (ed.), *Writing the Social Text: Poetics and Politics in Social Science Discourse* (pp. 73–90). New York : Aldine.

Levi-Strauss, C. (1973). *Totemism.* Harmondsworth, U.K.: Penguin.

Lindley, J., Bryan, S., and Conley, B. (1981). "Near-Death Experiences in a Pacific Northwest American Population: The Evergreen Study." *Anabiosis* 1: 104–24.

Lockley, R. M. (1976). *The Private Life of the Rabbit: An Account of the Life History and Social Behaviour of the Wild Rabbit.* London: Andre Deutsch.

Lorimer, D. (1990). *Whole in One: The Near-Death Experience and the Ethic of Interconnectedness.* London: Arkana.

Lucas, R. A. (1968). "Social Implications of the Immediacy of Death." *Canadian Review of Sociology and Anthropology* 5: 1–16.

Lundahl, C. R. (1981–82). "The Perceived Otherworld in Mormon Near-Death Experience: A Social and Physical Description." *Omega* 12: 319–27.

Lyotard, J. F. (1984). *The Postmodern Condition: A Report on Knowledge.* Manchester, U.K.: Manchester University Press.

Malino, J. R. (1966). "Coping with Death in Western Religious Civilisation." *Zygon: Journal of Religion and Science* 1: 354–65.

Mannheim, K. (1960). *Ideology and Utopia.* London: Routledge and Kegan Paul.

Manuel, F. E., and Manuel, F. R. (1979). *Utopian Thought in the Western World.* Oxford: Basil Blackwell.

Marshall, V. (1980). *Last Chapters: A Sociology of Aging and Dying.* Monterey, Calif.: Brooks/Cole.

Matchett, W. F. (1972). "Repeated Hallucinatory Experiences as a Part of the Mourning Process among Hopi Indian Women." *Psychiatry* 35: 185–94.

McNulty, F. (1992, 6 Dec.). "Children's Books for Christmas." *New Yorker:* 176–82.

Midgley, M. (1992). *Science as Salvation: A Modern Myth and Its Meaning.* London: Routledge.

Mills, A. (1991). "Commentary on Allan Kellehear's 'Near-Death Experiences and the Pursuit of the Ideal Society.'" *Journal of Near-Death Studies* 10: 113–22.

Mills, C. W. (1959). *The Sociological Imagination.* New York: Oxford University Press.

Mills, C. W. (1963). "Language, Logic and Culture." In I. L. Horowitz (ed.), *Power, Politics and People: The Collected Essays of C. Wright Mills* (pp. 423–38). New York: Oxford University Press.

Mitford, J. (1980). *The American Way of Death.* London: Quartet.

Moody, R. A., Jr. (1975). *Life after Life.* Covington, Ga.: Mockingbird.

Moody, R. A., Jr., with Perry, P. (1988). *The Light Beyond.* London: Macmillan.

Moody, R. A., Jr., and Perry, P. (1993). *Reunions: Visionary Encounters with Departed Loved Ones.* New York: Villard.

Morabito, L. (1990). "Love and God in the Near-Death Experience" (letter). *Journal of Near-Death Studies* 9: 65.

Morrice, J. K. W. (1976). *Crisis Intervention   Studies in Community Care* (p. 20). Oxford: Pergamon.

Morse, M. (1989). "Comments on 'A Neurobiological Model for Near-Death Experiences.'" *Journal of Near-Death Studies* 7: 224.

Morse, M., Castillo, P., Veneccia, D., Milstein, J., and Tyler, D. (1990). "Childhood Near-Death Experiences." *American Journal of Diseases of Children* 140: 1110–14.

Morse, M., Connor, P., and Tyler, D. (1985). "Near-Death Experiences in a Pediatric Population." *American Journal of Diseases of Children* 139: 595–600.

Morse, M., with Perry, P. (1990). *Closer to the Light.* London: Souvenir.

Naisbitt, J. (1984). *Megatrends.* London: MacDonald.

Neihardt, R. (1961). *Black Elk Speaks.* Lincoln: University of Nebraska Press.

Neppe, V. P. (1989). "Near-Death Experiences: A New Challenge in Temporal Lobe Phenomenology? Comments on 'A Neurobiological Model for Near-Death Experiences.'" *Journal of Near-Death Experiences* 7: 243–47.

Neumann, M. (1992). "The Trail Through Experience: Finding Self in the Recollection of Travel." In (pp. 176–201). C. Ellis and M. G. Flaherty (eds.), *Investigating Subjectivity: Research on Lived Experiences* Newbury Park, Calif.: Sage.

Nicholson, M., and McLaughlin, C. (1988). "Social Constructionism and Medical Sociology: A Study of the Vascular Theory of Multiple Sclerosis." *Sociology of Health and Illness* 10: 234–61.

Noyes, R. (1972). "The Experience of Dying." *Psychiatry* 35: 174–84.

Noyes, R. (1989). "Comments on 'A Neurobiological Model for Near-Death Experiences.'" *Journal of Near-Death Studies* 7: 249–50.

Noyes, R., and Kletti, R. (1976). "Depersonalization in the Face of Life Threatening Danger: A Description." *Psychiatry* 39: 1927.

Noyes, R., and Kletti, R. (1977) "Panoramic Memory: A Response to the Threat of Death." *Omega* 8: 181–94.

Oakes, A. (1984). "Near Death Events and Critical Care Nursing. In Bruce Greyson and Charles Flynn (eds.)," The Near-Death Experience: Problems, Prospects and Perspectives (pp. 223–31). Springfield, Ill.: Charles C Thomas.

Ogasawara, S. (1963). *Chugoku Kinsei Jodokyoshi no Kenkyu* (*Research on the History of Pure Land Buddhism in Recent China*). Kyoto: Hyakkaen.

Orne, R. (1986). "Nurse's Views of NDEs." *American Journal of Nursing* 86: 419–20.

Osis, K., and Haraldsson, E. (1977). *At the Hour of Death*. New York: Avon.

Oxford English Dictionary (1989). Oxford: Oxford University Press.

Panoff, M. (1968). "The Notion of the Double Self Among the Maenge." *Journal of the Polynesian Society* 77: 275–95.

Paraskos, J. (1992). "Biblical Accounts of Resuscitation." *Journal of the History of Medicine and Allied Sciences* 47: 310–21.

Pasricha, S. (1992). "Near-Death Experiences in South India. A Systematic Survey in Channapatua." *NIHMANS Journal* 10: 111–18.

Pasricha, S. (1993). "A Systematic Survey of Near-Death Experiences in South India." *Journal of Scientific Exploration* 7: 161–71.

Pasricha, S., and Stevenson, I. (1986). "Near-Death Experiences in India: A Preliminary Report." *Journal of Nervous and Mental Disease* 174: 165–70.

Peale, Norman Vincent. (1953). *The Power of Positive Thinking*. Surrey, U.K.: World's Work.

Perry, M. (1992). "Assessment of Clergy Knowledge and Attitudes" (letter). *Journal of Near-Death Studies* 11: 129.

Persinger, M. A. (1989). "Modern Neuroscience and Near-Death Experiences: Expectancies and Implications. Comments on 'A Neurobiological Model for Near-Death Experiences.'" *Journal of Near-Death Studies* 7: 233–39.

Personal Correspondence Department, Worldwide Church of God, Form *Letter from Pastor General Joseph W. Tkach,* undated.

Pfister, O. (1981) "Shock Thoughts and Fantasies in Extreme Mortal Danger." Translated by R. Kletti and R. Noyes, Jr., as "Mental States in Mortal Danger." *Essence* 5: 5–20.

Polanyi, M. (1973). *Personal Knowledge*. London: Routledge and Kegan Paul.

Popper, K. R. (1968). *The Logic of Scientific Discovery*. London: Hutchinson.

Powles, J. (1973). "On the Limitations of Modern Medicine." *Science, Medicine and Man* 1: 1–30.

Raphael, B. (1986). *When Disaster Strikes*. New York: Basic Books.

Readers Digest Association (1977). *Folklore, Myths and Legends of Britain*. London: Readers Digest Association.

Revitalized Signs: Newsletter for the International Association for Near-Death Studies, Vols. 8 and 9.

Ring, K. (1980). *Life at Death: A Scientific Investigation of the Near-Death Experience*. New York: Coward, McCann & Geohegan.

Ring, K. (1984). *Heading Toward Omega: In Search of the Meaning of the Near-Death Experience*. New York: William Morrow.

Ring, K. (1988). "Prophetic Visions in 1988: A Critical Reappraisal." *Journal of Near-Death Studies* 7: 4–18.

Ring, K. (1991). "Amazing Grace: The Near-Death Experience as Compensatory Gift." *Journal of Near-Death Studies* 10: 11–39.

Ritchie, G. (1978). *Return from Tomorrow*. Waco, Tex.: Chosen Books.

Roberts, G., and Owen, J. (1988). "The Near-Death Experience." *British Journal of Psychiatry* 153: 607–17.

Roberts, K. (1984). *Religion in Sociological Perspective*. Homeward, Ill.: Dorsey Press.

Roberts, R., and Kloss, R. (1979). *Social Movements*. London: C. V. Mosby.

Robertson, D. (1973). *Survive the Savage Sea*. London: Elek Books.

Roheim, G. (1932). "Psychoanalysis of Primitive Cultural Types." *International Journal of Psychoanalysis* 13 (1, 2): 1–224.

Rouhier, A. (1927). *La Plante qui fait les yeux emerveilles; le peyotl (Echinocactus Williamsii Lem)*. Paris: Doin.

Royse, D. (1985). "The Near Death Experience: A Survey of Clergy's Attitudes and Knowledge." *The Journal of Pastoral Care* 39: 31–42.

Rustin, M., and Rustin, M. (1987). *Narratives of Loss and Love: Studies in Modern Children's Fiction*. London: Verso.

Saavedra-Aguila, J. C., and Gomez-Jeria, J. S. (1989). "A Neurobiological Model for Near-Death Experiences." *Journal of Near-Death Studies* 7: 205–222 and rejoinder, 265–72.

Sabom, M. B. (1982). *Recollections of Death: A Medical Investigation*. New York: Harper and Row.

Safar, P. (1989). "Initiation of Closed Chest Cardiopulmonary Resuscitation Basic Life Support: A Personal History." *Resuscitation* 18: 7–20.

Sagan, C. (1979). *Broca's Brain: Reflections on the Romance of Science*. New York: Random House.

Schoolcraft, H. R. (1825). *Travels in the Central Portion of the Mississippi Valley*. New York: Collins and Henry.

Schorer, C. E. (1985). "Two Native North American Near-Death Experiences." *Omega* 16: 111–13.

Seigal, R. K. (1980). "The Psychology of Life after Death." *American Psychologist* 35: 911–31.

Seigal, R. K. (1981, Jan.). "Accounting for Afterlife Experiences." *Psychology Today*: 65–75.

Serdahely, W. J. (1989–90). "A Pediatric Near-Death Experience: Tunnel Variants." *Omega* 20: 55–62.

Serdahely, W. J. (1990). "Pediatric Near-Death Experiences." *Journal of Near-Death Studies* 9: 33–39.

Shiels, D. (1978). "A Cross-Cultural Study of Beliefs in Out-of-the-Body Experiences, Waking and Sleeping." *Journal of the Society for Psychical Research* 49: 697–741.

Simon-Buller, S., Christopherson, V. A., and Jones, R. A. (1988–89). "Correlates of Sensing the Presence of a Deceased Spouse." *Omega* 19: 21–30.

Simpson, M. (1979). *Dying, Death and Grief: A Critically Annotated Bibliography and Source Book of Thanatology and Terminal Care*. New York: Plenum Press.

Smelser, N. (1963). *Theory of Collective Behavior*. New York: Free Press.

Starr, P. (1982). *The Social Transformation of American Medicine*. New York: Basic Books.

Stevenson, I., Cook, E. W., and McClean-Rice, N. (1989–90). "Are Persons Reporting 'Near-death Experiences' Really Near Death? A Study of Medical Records." *Omega* 20: 45–54.

Stott, J. C. (1984). *Children's Literature from A to Z: A Guide for Parents and Teachers*. New York: McGraw-Hill.

Straight, S. (1984). "A Wave among Wave: Katherine Anne Porter's Near-Death Experience." *Anabiosis* 4: 107–23.

Suedfeld, P., and Mocellin, J. S. P. (1987). "The 'Sensed Presence' in Unusual Environments." *Environment and Behavior*. 19: 33–52.

Sutherland, C. (1990). "Changes in Religious Beliefs, Attitudes and Practices Following Near-Death Experiences: An Australian Study." *Journal of Near-Death Studies* 9: 21–31.

Sutherland, C. (1992). *Transformed by the Light*. Sydney: Bantam.

Taylor, R. (1979). *Medicine Out of Control*. Melbourne: Sun Books.

The Age (1989 7 Oct.). "My Tale, by the Yacht's Skipper": 1, 4.

Tolstoy, L. (1960). *The Death of Ivan Ilyich and Other Stories*. New York: New American Library (Signet).

Toulmin, S. (1990). *Cosmopolis: The Hidden Agenda of Modernity*. New York: Free Press.

Turner, B. S. (1992). *Regulating Bodies: Essays in Medical Sociology*. London: Routledge.

Varon, J., and Sternbach, G. (1991). "Cardiopulmonary Resuscitation: Lessons from the Past." *Journal of Emergency Medicine* 9: 503–7.

Veatch, R. (1976). *Death, Dying and the Biological Revolution*. New Haven, Conn.: Yale University Press.

Veatch, R. M., and Tai, E. (1980). "Talking about Death: Patterns of Lay and Professional Change." *American Academy of Political and Social Sciences* 447: 29–45.

Vrtis, M. (1992). "Cost/Benefit Analysis of Cardiopulmonary Resuscitation: A History of CPR—Part 1." *Nursing Management* 23: 50–54.

Walker, B., and Russell, R. (1989). "Assessing Psychologists' Knowledge and Attitudes Toward Near-Death Phenomena." *Journal of Near-Death Studies* 8: 103–10.

Walter, T. (1993). "Death in the New Age." *Religion* 23: 127–45.

Walters, K. S. (1989). *The Sane Society Ideal in Modern Utopianism*. Toronto: Edwin Mellen Press.

Warner, W. L. (1937). *A Black Civilization: A Social Study of an Australian Tribe*. New York: Harper and Brothers.

Weales, G. (1983, 3 Apr.). "Children's Books." *New York Times Book Review:* 13.

Weber, M. (1947). *The Theory of Social and Economic Organization*. New York: Free Press.

Weber, M. (1965). *The Sociology of Religion*. London: Methuen.

Weibust, P. (1991). "Being One with God Is Something That Can Be Done Without Rules: Commentary on Allan Kellehear's 'Near-Death Experiences and the Pursuit of the Ideal Society.' " *Journal of Near-Death Studies* 10: 107–11.

White, K., and Willis, E. (1992). "The Languages of AIDS." *New Zealand Sociology* 7: 127–49.

Williams, M. (1922). *The Velveteen Rabbit: or How Toys Become Real*. New York: Doubleday.

Willis, E. (1989). *Medical Dominance*. Sydney: Allen & Unwin.

Wren-Lewis, J. (1986). "Joy Without a Cause: An Anticipation of Modern 'Near-Death Experiences' Research in G. K. Chesterton's Novel *The Ball and the Cross*." *The Chesterton Review* 12: 49–61.

Zaleski, C. (1987). *Otherworld Journeys: Accounts of Near-Death Experiences in Medieval and Modern Times*. New York: Oxford University Press.

Zhi-ying, F., and Jian-Xun, L. (1992). "Near-Death Experiences among Survivors of the 1976 Tangshan Earthquake." *Journal of Near-Death Studies* 11: 39–48.

Zola, I. K. (1972). "Medicine as an Institution of Social Control." *Sociological Review* 20: 487–504.

NAME INDEX

SUBJECT INDEX